Intermediate Algebra

Concepts Through Applications
Class Test Edition, volume 2

MARK CLARK
Palomar College

BROOKS/COLE
CENGAGE Learning™

Australia • Brazil • Japan • Korea • Mexico • Singapore • Spain • United Kingdom • United States

BROOKS/COLE
CENGAGE Learning™

Intermediate Algebra : Concepts through Applications Class Test Edition, vol. 2
Mark Clark

Executive Editor: Charlie Van Wagner

Development Editor: Donald Gecewicz

Assistant Editor: Shaun Williams

Editorial Assistant: Mary De La Cruz

Marketing Manager: Joe Rogove

Marketing Assistant: Angela Kim

Project Manager, Editorial Production: Cheryll Linthicum

Creative Director: Rob Hugel

Art Director: Vernon T. Boes

Photo Researcher: Terri Wright

Copy Editor: Barbara Willette

Illustrators: Hearthside, ICC India

Print Buyer: Barbara Britton

Permissions Editor: Stephanie Lee

Production Service: Hearthside Publishing Services

Text Designer: Geri Davis

Cover Designer: Roger Knox

Cover Image: Photonica

Compositor: ICC Macmillan Inc.

For product information and technology assistance, contact us at **Cengage Learning Customer & Sales Support, 1-800-354-9706.**

For permission to use material from this text or product, submit all requests online at **www.cengage.com/permissions**
Further permissions questions can be e-mailed to **permissionrequest@cengage.com**

Student Edition:

ISBN-13: 978-0-495-82843-3

ISBN-10: 0-495-82843-2

Brooks/Cole Cengage Learning
10 Davis Drive
Belmont, CA 94002-3098
USA

Cengage Learning is a leading provider of customized learning solutions with office locations around the globe, including Singapore, the United Kingdom, Australia, Mexico, Brazil, and Japan. Locate your local office at **international.cengage.com/region**.

Cengage Learning products are represented in Canada by Nelson Education, Ltd.

For your course and learning solutions, visit **academic.cengage.com**

Purchase any of our products at your local college store or at our preferred online store **www.ichapters.com**

Printed in the United States of America
1 2 3 4 5 6 7 12 11 10 09 08

Contents

7 RATIONAL FUNCTIONS

8 RADICAL FUNCTIONS

Exponential Functions

- Recognize an exponential growth or decay pattern from data or context.

- Find an exponential model using a pattern.

- Solve power equations using exponent rules.

- Understand the relationship between the parameters of an exponential equation and the graph.

- Sketch the graph of an exponential function by hand.

- Find and use an exponential model when appropriate.

- Understand and find the domain and range of an exponential function.

- Understand growth rates and their relationship to exponential functions.

- Understand and use compounding interest formulas.

© Peter Arnold, Inc./Alamy

ow could such a small animal cause so much damage? Rabbits like the ones pictured here cause an estimated $600 million in losses to Australian farmers every year. Rabbits eat vegetation that other animals could feed on, endangering sheep and cattle and contributing to the extinction of many native plants and animal species. In this chapter we will study exponential growth and decay patterns and how to model these situations mathematically. One of the chapter projects will ask you to use these functions to investigate the exponential growth of the Australian rabbit population.

  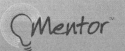

Introduction to Exponentials

EXPLORING EXPONENTIAL GROWTH AND DECAY

In this chapter we will add yet another type of function to our list of possible models. So far we have found linear and quadratic models and have worked with systems of equations. Now we will discuss another function that is found in many areas of science and life called an exponential function.

© Ted Horowitz/CORBIS

The mathematics of uncontrolled growth are frightening. A single cell of the bacterium E. coli would, under ideal circumstances, divide every twenty minutes. That is not particularly disturbing until you think about it, but the fact is that bacteria multiply geometrically: one becomes two, two become four, four become eight, and so on. In this way it can be shown that in a single day, one cell of E. coli could produce a super-colony equal in size and weight to the entire planet Earth.

Michael Crichton (1969) *The Andromeda Strain* (New York: 1969), p. 247.

This quote might sound unrealistic. However, with the assumption of "ideal circumstances" the mathematics of this type of situation is amazing. Although Crichton calls this type of growth *geometric,* today it is more often called *exponential growth.* Later in this section we will investigate this situation and see just how fast *E. coli* bacteria grows in "ideal circumstances." First let's consider a simple version of this situation.

CONCEPT INVESTIGATION I HOW FAST ARE THOSE BACTERIA GROWING? ·····································

Bacterial growth involves the process of doubling and then doubling again and again. Let's assume that we are doing an experiment in a biology lab starting with a single bacteria cell that in the current environment will split into two new cells every hour.

a. Complete the table below with the number of bacteria, *B*, after *h* hours. Remember that the number of bacteria is doubling every hour.

Hours	Bacteria
0	1
1	2
2	4
3	
4	
5	
6	

b. Create a scatterplot for these data on your calculator. Describe the shape of the graph.

c. Now rewrite the table from part a, writing out the calculations, not the final results.

Hours	Bacteria
0	1
1	$1 \cdot 2$
2	$1 \cdot 2 \cdot 2$
3	
4	
5	
6	

d. As you can see from this last table, this situation is calculated by using repeated multiplication. Recall that exponents are a shorter way of representing repeated multiplication. Use exponents to rewrite this table and find a pattern for a model that will give the number of bacteria after *h* hours.

Hours	Bacteria
0	1
1	$1(2)$
2	$1(2)^2$
3	$1(2)^3$
4	
5	
6	
h	

e. Graph your model with the data. How well does your model fit the data?

f. Use your model to find the number of bacteria after 12 hours and 24 hours.

⬥

An **exponential function** is based on a pattern of repeated multiplication that leads to using exponents to simplify the expression. Because the exponent in this type of model is what changes from one input value to the other, you end up with a variable in the exponent rather than a constant. It is this variable in the exponent that makes the function exponential rather than a parabola or other type of function.

> **DEFINITIONS**
> **Exponential function:** A basic exponential function can be written in the form
> $$f(x) = a \cdot b^x$$
> where a and b are real numbers, $a \neq 0$, $b > 0$, and $b \neq 1$.
> **Base:** The constant b is called the base of an exponential function.

Finding a model for exponential data can be done in several ways. However, a common approach involves using a situation and looking for a pattern. This will take practice and, in some cases, a new way of thinking. Read through the following examples and problems. Paying close attention to how the final pattern is found that leads to a model.

EXAMPLE 1 EXPONENTIAL GROWTH PATTERN

A rumor is spreading across a local college campus that there will be no finals for any classes this summer. At 8 A.M. today, seven people have heard the rumor. Assume that after each hour, 3 times as many students have heard the rumor. For example, 21 people have heard the rumor by 9 A.M. Let $R(t)$ represent the number of people who have heard the rumor t hours after 8 A.M.

a. Find a formula for $R(t)$.

b. If the college has about 15,000 students, estimate numerically the time at which all of them would have heard the rumor.

c. How many people have heard the rumor by 8 P.M. that night?

Solution

a. First we need to define the variables and to create a table.

$R(t)$ = the number of people who have heard the rumor

t = time in hours after 8 A.M.

We start with seven people at 8 A.M.. Each hour, 3 times as many people have heard the rumor. We can start building the table by writing out the calculations and then simplify later.

t	$R(t)$	
0	7	Start with 7 people.
1	$7 \cdot 3$	One hour later 3 times as many people.
2	$7 \cdot 3 \cdot 3$	Another hour later and 3 times as many people.
3	$7 \cdot 3 \cdot 3 \cdot 3$	
4	$7 \cdot 3 \cdot 3 \cdot 3 \cdot 3$	Every hours after, 3 times as many people have heard the rumor.
5	$7 \cdot 3 \cdot 3 \cdot 3 \cdot 3 \cdot 3$	
6	$7 \cdot 3 \cdot 3 \cdot 3 \cdot 3 \cdot 3 \cdot 3$	

Because each hour 3 times as many people have heard the rumor, we have repeated multiplication by 3. Each row in the table can be simplified using exponents.

t	$R(t)$
0	7
1	$7(3)$
2	$7(3)^2$
3	$7(3)^3$
4	$7(3)^4$
5	$7(3)^5$
6	$7(3)^6$

In looking at this pattern, notice that the 7 and 3 are the same in each expression, after the first. The only part of each expression that is changing is the exponent itself. Note that the input value that t has for each row is the same as the exponent in the resulting expression.

t	$R(t)$
0	7
	$1 = 1$
1	$7(3)^1$
	$2 = 2$
2	$7(3)^2$
	$3 = 3$
3	$7(3)^3$
	$4 = 4$
4	$7(3)^4$
	$5 = 5$
5	$7(3)^5$
	$6 = 6$
6	$7(3)^6$

Because the input value is the same as the exponent in each row of the table, if we consider t to be the input, the exponent will also be t. With this in mind, our model can be expressed by, $R(t) = 7(3)^t$.

If we graph these data and our model, we get the following.

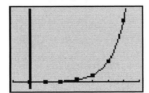 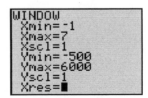

b. Using the table we get the following.

X	Y1
4	567
5	1701
6	5103
7	15309
8	45927

Y1⊟7*3^X

The model reaches 15,309 people having heard the rumor when $t = 7$, so by 3 P.M. all of the students at the college should have heard the rumor.

c. 8 P.M. that night is 12 hours after 8 A.M., so we can substitute $t = 12$ into our model.

$$R(12) = 7(3)^{12}$$
$$R(12) = 3720087$$

This means that 3,720,087 people will have heard this rumor by 8 P.M. that night. This is clearly model breakdown, since 3.7 million people will not have heard this rumor in 12 hours.

This exponential model is growing so quickly that its domain must be carefully determined to avoid obvious model breakdown. We will discuss the domain for an exponential function in Section 5.3.

EXAMPLE ① PRACTICE PROBLEM

A loan shark lends desperate people money at a very high rate of interest. You have found yourself in trouble and need to borrow $5000 to pay your legal bills. One loan shark will lend you the money under the rules stated in the I Owe You shown below.

IOU
Original loan amount $5000.00
Each week the balance will double until paid in full

a. Find an equation for a model that will give the balance on this loan after w weeks have gone by.

b. Find the balance on this loan if you keep the money for 4 weeks.

c. Estimate numerically after how many weeks the balance of the loan will exceed 1 million dollars.

Now let's look at the *E. coli* example with which we started the section.

EXAMPLE 2 EXPONENTIAL GROWTH PATTERN

The bacteria *E. coli* will double every 20 minutes under "ideal circumstances." If we assume these conditions and start with one bacterium, find the following.

a.   Find an equation for a model for the number of *E. coli* bacteria.

b.   Use your model to determine the number of bacteria after 12 hours.

c.   Use your model to determine the number of bacteria after 24 hours.

Solution

There are several ways in which you could define your variables and thus get different models for this context. In this problem we will consider two of those options.
   Option 1: h = 1 represents one hour.

a.   Let's start by defining some variables and creating a table of data. One decision that we need to make is what unit of time we should use for this model. The most common time measurement might be hours, but because the number of bacteria doubles every 20 minutes, a 20-minute interval may be used as well. We will do this problem both ways and see how the models compare.

   E = the number of *E. coli* bacteria

   h = time in hours since the beginning of the experiment

Now let's fill in the table. Remember that the number of bacteria is doubling every 20 minutes, not every hour. This means that the number of bacteria will double 3 times each hour.

h (hours)	E(h)
0	1
1	$1 \cdot 2 \cdot 2 \cdot 2 = 1(2)^3$
2	$1 \cdot 2 \cdot 2 \cdot 2 \cdot 2 \cdot 2 \cdot 2 = 1(2)^6$
3	$1 \cdot 2 \cdot 2 \cdot 2 \cdot 2 \cdot 2 \cdot 2 \cdot 2 \cdot 2 \cdot 2 = 1(2)^9$
4	$1(2)^{12}$

In looking for a pattern, be very careful to make sure that your model gives you the results you want. In this table we see that the 1 and 2 are the same for each exponential expression. Only the exponent is changing. We need to find a relationship between the input values of *t* and the exponent that is changing. If you look carefully, you should see that the exponent is 3 times as much as the input value.

The pattern in the table shows that we need to multiply the input value by 3 to get the exponent. Doing so leads to the model $E(h) = 1(2)^{3h}$. If we check this model by graphing, it gives us the expected results.

b. $h = 12$ represents 12 hours later, so we substitute 12 into our model and get

$$E(12) = 1(2)^{3(12)}$$
$$E(12) = 1(2)^{36}$$
$$E(12) = 68719476736$$

So after only 12 hours there will be 68.7 billion *E. coli* bacteria.

c. The time $h = 24$ represents 24 hours later, so we substitute 24 into our model and get

$$E(24) = 1(2)^{3(24)}$$
$$E(24) = 1(2)^{72}$$
$$E(24) = 4.722 \times 10^{21} \quad \text{This answer is given in scientific notation.}$$

This means that there will be 4.7 billion trillion *E. coli* bacteria after 24 hours. Now that's a lot of bacteria! This could happen only in "ideal circumstances" and is therefore most likely model breakdown, but the amazingly rapid growth is still true earlier in the process.

Solution   *Option 2: $n = 1$ represents 20 minutes.*

a. If we define the time as the number of 20-minute intervals that pass from the beginning of the experiment, we would change our table and model accordingly. Let's redefine the variables as

$E =$ the number of *E. coli* bacteria

$n =$ the number of 20-minute intervals since the beginning of the experiment

Now let's fill in the table, remembering that $n = 1$ represents only 20 minutes, so the bacteria will have doubled once.

n (20-minute periods)	$E(n)$
0	1
1	$1 \cdot 2 = 1(2)$
2	$1 \cdot 2 \cdot 2 = 1(2)^2$
3	$1 \cdot 2 \cdot 2 \cdot 2 = 1(2)^3$
4	$1(2)^4$
5	$1(2)^5$
6	$1(2)^6$

This pattern is the same as the one in the first example that we did in this section and results in the model $E(n) = 1(2)^n$. Although this model is the same as the one in Example 1, we will have to interpret it very carefully to get the correct results.

b. With the new input variable n, 12 hours will have 36 intervals of 20 minutes, 3 per hour. Therefore 12 hours will be represented by $n = 36$. This results in

$$E(36) = 1(2)^{36}$$
$$E(36) = 68719476736$$

This is the same result as the previous model: In 12 hours there will be 68.7 billion bacteria.

c. This time $n = 72$ represents 24 hours later, so we substitute 72 into our model and get

$$E(72) = 1(2)^{72}$$
$$E(72) = 4.722 \times 10^{21} \quad \text{This answer is given in scientific notation.}$$

This means that there will be 4.7 billion trillion *E. coli* bacteria after 24 hours, as in the previous model.

As you can see, there were two ways to look at this pattern. Most people are comfortable with measuring time in hours, but few are comfortable with measuring time in 20-minute intervals. Neither of these models is better than the other, but you must use caution when using and interpreting your models. Please note that these are not the only two possibilities, but they are two common ones.

EXAMPLE ② PRACTICE PROBLEM

Suppose that under "ideal conditions" human beings could double their population size every 50 years and that there were 6 billion humans on the earth in the year 2000.

a. Find an equation for a model for the world population.

b. Estimate the world population in 2500.

c. Estimate graphically when the world population would reach 10 billion.

Population Density Map of the World The brightest areas are the most densely populated.

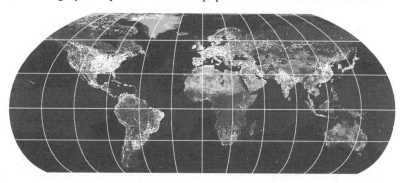

So far we have seen several examples of **exponential growth.** The opposite of this type of growth is **exponential decay.** Exponential decay plays a large role in sciences such as archeology, in which scientists use exponential decay of carbon-14 to date historical objects. The exponential decay of radioactive elements is also a large part of the concerns people have with nuclear power plants. The basic concept driving most exponential decay problems is the **half-life** of

Say What?

There are many words or phrases that imply that a situation will be modeled by an exponential function. Here are just a few of them you may see:

- Exponential growth
- Exponential decay
- Half-life
- Growth or decay by a factor of:
    fraction
    percentage
- Double every . . .
- Triple every . . .
- Repeated multiplication

the element. The half-life is simply a measurement of how long it takes before only half of an initial quantity still remains. The half-lives of different elements have large ranges of values, such as radon-222, which has a half-life of only 3.825 days, and rubidium-87 with a half-life of 49 billion years. To get a better idea of how this works, let's look at a few examples and problems dealing with exponential decay.

EXAMPLE 3 EXPONENTIAL DECAY - HALF LIFE

Carbon-14 is an isotope that is found in all living creatures. Once a creature dies and stops taking in new carbon, the existing carbon-14 decays and is no longer replaced. Because carbon-14 has a half-life of 5700 years, we can use the amount of carbon-14 that is left in an artifact to determine the age of the formerly living thing. Let's assume that an artifact, such as a carving made of wood, that has been dead for some time started with 300 atoms of carbon-14 at the time of death.

a. Find an equation for a model for the amount of carbon-14 remaining in the artifact.

b. Estimate the amount of carbon-14 that should be in the artifact if it is 34,200 years old.

Solution

a. We know that the original amount of carbon-14 was 300 atoms and that half of that amount will be left after 5700 years. With this information we define variables and create a table to help us find a pattern.

C = the amount of carbon-14 atoms in the artifact

t = time in years since the artifact died

We do not know what happens after one year. We do know that after 5700 years half of the carbon-14 remains. Because we only know what happens every 5700 years we will count the years by 5700's.

t	$C(t)$		t	$C(t)$
0	300		0	300
5700	$300\left(\dfrac{1}{2}\right)$		5700	$300\left(\dfrac{1}{2}\right)$
11,400	$300\left(\dfrac{1}{2}\right)\left(\dfrac{1}{2}\right)$		11,400	$300\left(\dfrac{1}{2}\right)^2$
17,100	$300\left(\dfrac{1}{2}\right)\left(\dfrac{1}{2}\right)\left(\dfrac{1}{2}\right)$		17,100	$300\left(\dfrac{1}{2}\right)^3$

Just as in the previous problems, the constant 300 and the base $\dfrac{1}{2}$ are the same in each row of the table. The only change is in the exponent. This means that we need to find the connection between the input values and the exponents. If you look carefully, you should note that the inputs are multiples of 5700. If you divide each input by 5700, you will get the exponent on that row.

t	$C(t)$
0	300
5700	$300\left(\dfrac{1}{2}\right)^1$
11,400	$300\left(\dfrac{1}{2}\right)^2$
17,100	$300\left(\dfrac{1}{2}\right)^3$

$5700 \div 5700 = 1$

$11400 \div 5700 = 2$

$17100 \div 5700 = 3$

If we let t represent the input value and divide it by 5700, we get the model

$$C(t) = 300\left(\frac{1}{2}\right)^{t/5700}$$

b.   34,200 years old can be represented by $t = 34200$, so we get

$$C(34200) = 300\left(\frac{1}{2}\right)^{34200/5700}$$

$$C(34200) = 300\left(\frac{1}{2}\right)^{6}$$

$$C(34200) = 4.6875$$

This means that after 34,200 years the artifact should have only about 4 or 5 atoms of carbon-14 remaining.

EXAMPLE PRACTICE PROBLEM

Uranium-238 is a radioactive element that occurs naturally in most types of granite and soil in varying degrees. As it undergoes radioactive decay, a chain of elements is formed. As each element decays it gives off another element until it reaches a stable element. Through this radioactive decay uranium-238 decays through several elements until it changes into stable, nonradioactive lead-206. Thorium-234 is the first by-product of the decay of uranium-238 and has a half-life of 24.5 days.

a.   Find an equation for a model for the percent of a thorium-234 sample left after d days.

b.   Estimate the percent of thorium-234 left after 180 days.

c.   Estimate numerically how many days there will be before only 40% of the sample is left.

EXAMPLE   **4** EXPONENTIAL DECAY PATTERN

The brightness of light is measured with a unit called a lumen. Sharp makes a high-end Conference Series XG-P25X LCD Projector that has a brightness of 4000 lumens. Using a series of several mirrors that reflect only $\frac{3}{5}$ of the light that hits it, a stage technician is trying to project a series of photos on to several walls of a concert stage. The technician is concerned with the brightness of the light that will remain after using several mirrors to place the projected photos in the right places.

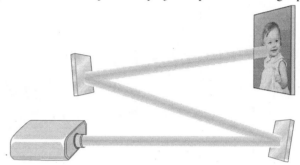

a. Find an equation for a model that will tell the technician the remaining lumens that will be projected after m mirrors have reflected the image.

b. Use your model to determine the lumens remaining after five mirrors.

c. If the technician knows that he needs a minimum of 500 lumens to have a good image, use the table to find the maximum number of mirrors he can use with this projector.

Solution

a. The projector will initially project a light with 4000 lumens, and after each mirror reflects the light, only $\frac{3}{5}$ of that light will remain. Let's define the following variables.

$L =$ the lumens of light remaining

$m =$ the number of mirrors used to reflect the light

m	$L(m)$
0	4000
1	$4000\left(\frac{3}{5}\right)$
2	$4000\left(\frac{3}{5}\right)^2$
3	$4000\left(\frac{3}{5}\right)^3$
m	$4000\left(\frac{3}{5}\right)^m$

This pattern results in the model

$$L(m) = 4000\left(\frac{3}{5}\right)^m$$

b.   Five mirrors is represented by $m = 5$, so we get

$$L(5) = 4000\left(\frac{3}{5}\right)^5$$

$$L(5) = 311.04$$

If the technician uses five mirrors to place the photo projection, it will have a brightness of only 311.04 lumens.

c.   Using table feature on the calculator, we get

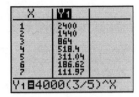

From this table we see that the technician can use only up to four mirrors to have the 500 lumens necessary to project a good image.

RECOGNIZING EXPONENTIAL PATTERNS

Whenever you are given data without a situation to consider, you will need to remember the basic concept of an exponential model to find a pattern. This fundamental idea is repeated multiplication by the same number, the base. When investigating data, look for the initial amount and the base.

EXAMPLE   **5**   FINDING AN EXPONENTIAL MODEL GIVEN DATA

Use the following tables to find exponential models of the given data.

a.

x	f(x)
0	15
1	60
2	240
3	960
4	3840

b.

x	f(x)
0	16
1	36
2	81
3	182.25
4	410.06

c.

t	h(t)
0	1500
4	300
8	60
12	12
16	2.4

Solution

a. For each of these tables we are given the initial value when the input variable is zero. This initial value is the value of a in the exponential model. We can find the base by dividing each consecutive output value by the previous output value.

x	f(x)	base
0	15	
		$\frac{60}{15} = 4$
1	60	
		$\frac{240}{60} = 4$
2	240	
		$\frac{960}{240} = 4$
3	960	
		$\frac{3840}{960} = 4$
4	3840	

As you can see from performing these divisions, the common multiplier is 4 for each row. If we rewrite each output showing how to calculate it using the base 4, we should see the pattern and therefore the appropriate model.

x	f(x)	base
0	15	15
1	60	$15(4)$
2	240	$15(4)^2$
3	960	$15(4)^3$
4	3840	$15(4)^4$

This leads us to the model $f(x) = 15(4)^x$.

b.

x	f(x)	base
0	16	
		$\frac{36}{16} = 2.25$
1	36	
		$\frac{81}{36} = 2.25$
2	81	
		$\frac{182.25}{81} = 2.25$
3	182.25	
		$\frac{410.0625}{182.25} = 2.25$
4	410.0625	

Again, by looking for the common multiplier, we have found a base. Because the initial value when $x = 0$ is 16, we get the model $f(x) = 16(2.25)^x$.

c. We really need to use caution when looking at this example, since the inputs are not one unit apart. Notice that each input is increasing by four units. This will change the pattern we find for this model.

t	h(t)	base
0	1500	
		$\frac{300}{1500} = 0.2$
4	300	
		$\frac{60}{300} = 0.2$
8	60	
		$\frac{12}{60} = 0.2$
12	12	
		$\frac{2.4}{12} = 0.2$
16	2.4	

Again, by looking for the common multiplier, we have found a base. Because the initial value when $t = 0$ is 1500, we know the constant. To find a pattern, it would be helpful to build another table showing the calculations.

t	h(t)
0	1500
4	1500(0.2)
8	$1500(0.2)^2$
12	$1500(0.2)^3$
16	$1500(0.2)^4$

As you can see, the inputs are not the same as the exponents. We need to find a pattern to show how to relate the input values to the changing exponents. You should be able to see that the exponents are the input values divided by 4. With this fact we get the model $h(t) = 1500(0.2)^{t/4}$.

EXAMPLE 5 PRACTICE PROBLEM

Use the following tables to find exponential models of the given data.

a.

x	f(x)
0	10
1	31
2	96.1
3	297.91
4	923.521

b.

x	f(x)
0	16
3	14.4
6	12.96
9	11.664
12	10.4976

5.1 Exercises

1. Under ideal circumstances a certain type of bacterium can double every hour. If a sample of 30 bacteria is allowed to grow in these ideal circumstances, answer the following.

  a. Find an equation for a model for the number of bacteria after h hours have passed.

  b. Estimate the number of bacteria present after 12 hours.

  c. Estimate the number of bacteria present after 24 hours.

  d. Estimate numerically when the number of bacteria will reach 1 million.

2. Under ideal circumstances a certain type of bacterium can triple every hour. If a sample of 20 bacteria is allowed to grow in these ideal circumstances, answer the following.

  a. Find an equation for a model for the number of bacteria after h hours have passed.

  b. Estimate the number of bacteria present after 5 hours.

  c. Estimate the number of bacteria present after 10 hours.

  d. Estimate numerically when the number of bacteria will reach 1 million.

3. Under ideal circumstances a certain type of bacterium can double every 15 minutes. If a sample of 3 bacteria is allowed to grow in ideal circumstances answer the following.

   a. Find an equation for a model for the number of bacteria after h hours have passed.

   b. Estimate the number of bacteria present after 5 hours.

   c. Estimate graphically when the number of bacteria will reach 1 million.

4. Under ideal circumstances a certain type of bacterium can triple every 20 minutes. If a sample of 5 bacteria is allowed to grow in ideal circumstances answer the following.

   a. Find an equation for a model for the number of bacteria after h hours have passed.

   b. Estimate the number of bacteria present after 7 hours.

   c. Estimate graphically when the number of bacteria will reach 1 million.

5. Under ideal circumstances a certain type of bacterium can double every 15 minutes. If a sample of 8 of these bacteria is allowed to grow in these ideal circumstances answer the following.

   a. Find an equation for a model for the number of bacteria after n 15-minute intervals have passed.

   b. Estimate the number of bacteria present after 3 hours.

   c. How do the models you found in exercises 3 and 5 compare?

6. Under ideal circumstances a certain type of bacterium can double every 20 minutes. If a sample of 45 of these bacteria is allowed to grow in these ideal circumstances answer the following.

   a. Find an equation for a model for the number of bacteria after n 20-minute intervals have passed.

   b. Estimate the number of bacteria present after 7 hours.

   c. How do the models you found in exercises 4 and 6 compare?

7. Under ideal circumstances a certain type of bacterium can triple every 30 minutes. If a sample of 8 of these bacteria is allowed to grow in these ideal circumstances answer the following.

   a. Find an equation for a model for the number of bacteria after h hours have passed.

   b. Estimate the number of bacteria present after 10 hours.

8. Under ideal circumstances a certain type of bacterium can double every 10 minutes. If a sample of 15 of these bacteria is allowed to grow in these ideal circumstances answer the following.

   a. Find an equation for a model for the number of bacteria after h hours have passed.

   b. Estimate the number of bacteria present after 4 hours.

9. Under ideal circumstances a certain type of bacterium can triple every 30 minutes. If a sample of 12 of these bacteria is allowed to grow in these ideal circumstances answer the following.

   a. Find an equation for a model for the number of bacteria after m minutes have passed.

   b. Estimate the number of bacteria present after 3 hours.

10. Under ideal circumstances a certain type of bacterium can double every 10 minutes. If a sample of 9 of these bacteria is allowed to grow in these ideal circumstances answer the following.

    a. Find an equation for a model for the number of bacteria after m minutes have passed.

    b. Estimate the number of bacteria present after 3 hours.

11. Under ideal circumstances a certain type of bacterium can triple every 30 minutes. If a sample of 10 of these bacteria is allowed to grow in these ideal circumstances answer the following.

    a. Find an equation for a model for the number of bacteria after n 30-minute intervals have passed.

    b. Estimate the number of bacteria present after 2 hours.

    c. How do the models you found in Exercises 7, 9 and 11 compare?

12. Under ideal circumstances a certain type of bacterium can double every 10 minutes. If a sample of 14 of these bacteria is allowed to grow in these ideal circumstances answer the following.

    a. Find an equation for a model for the number of bacteria after n 10-minute intervals have passed.

    b. Estimate the number of bacteria present after 6 hours.

    c. How do the models you found in Exercises 8, 10 and 12 compare?

13. The U.S. Census Bureau reported that the number of centenarians (people older than 100) is increasing rapidly. According to the report, the number of Hispanic centenarians is doubling about every 7.5 years. In 1990 there were approximately 2072 Hispanic centenarians.

Source: National Institute on Aging Journal, June 1999.

    a. If this trend continues, find an equation for a model that would predict the number of Hispanic centenarians.

    b. Use your model to estimate the number of Hispanic centenarians in 2050.

14. According to the same U.S. Census report discussed in Exercise 13, the number of centenarians in general will double every 10 years. In 1990 there were about 37,000 centenarians in the U.S.

Source: National Institute on Aging Journal, June 1999.

    a. If this trend continues, find an equation for a model that would predict the number of centenarians.

    b. Use your model to estimate the number of centenarians in 2050.

    c. According to census estimates, the number of centenarians in 2050 could be as high as

4.2 million people. How does your estimate compare?

Use this article to answer Questions 15 and 16.

POPULATION NEWS

India: Population is expected to double every 33 years. In 1971 there were about 560 million people in India. In 2000 India's population broke 1 billion.

Delhi: Is a territory in India and is one of the fastest-growing areas in India. The Delhi territory's pop-ulation has doubled about every 20 years since 1950.

Source: Census India 2001, CensusIndia.net.

15. a. Using this information, find an equation for a model for the population of India.

    b. Using your model, estimate the population of India in 2015.

    c. Estimate graphically when India's population will reach 2 billion.

16. a. Using this information, find an equation for a model for the population of Delhi, India if in 1950 there were approximately 2.2 million people.

    b. Using your model, estimate the population of Delhi, India in 2005.

    c. In 2001 the population of Delhi was about 13.8 million people. How does this compare with your projections for 2005?

17. Internet Search: Use the Internet to find how long people project it will take the U.S. population to double.

18. Internet Search: Use the Internet to find how long people project it will take the population of France to double.

19. Suppose you are given the two different options presented in the offer letter.

> Joe Smith Associates
> 123 Bank Ave.
> Dallas, TX 50344
>
>
> Dear Sir,
>
> We are offering two salary options at this time.
>
> **Option 1:** $2000 each week.
>
> **Option 2:** You earn 1 penny the first week, 2 pennies the next week, and so on, doubling your salary each week.
>
> The job is going to last for 25 weeks.
>
> Thank you,
>
> *Joe L. Smith*

a. Find the total salary for the 25 weeks if you take option 1.

b. Before you calculate the total for option 2, decide which of the two options you would want at this point.

c. Find an equation for a model for the salary you will earn in week w of this temporary job under option 2.

d. Find the total amount of salary you will earn using option 2. (Note that you will need to add up all 25 weeks' salary.)

e. Now which salary option would you choose for this job?

f. What option would be best for a 20-week job?

20. You are given the following two salary options for a 20 week temporary job.

Option 1: You earn $5000 the first week, $6000 the second week, and so on, adding $1000 to your salary each week.

Option 2: You earn 1 penny the first week, 3 pennies the next week, and so on, tripling your salary each week.

a. Find an equation for a model for the salary you will earn in week w of this temporary job under option 1. (Note that you are adding an amount not multiplying so your model will be linear.)

b. Find the total salary for the 20 weeks if you take option 1.

c. Find an equation for a model for the salary you will earn in week w of this temporary job under option 2.

d. Find the total amount of salary you will earn using option 2. (Note that you will need to add up all 20 weeks' salary.)

e. Which salary option would you choose for this job?

f. What option would be best for a 15-week job?

21. Recent advancement in the forensic dating of skeletal remains has led forensic pathologists to monitor the decay of certain isotopes found in humans. The isotope lead-210 has a half-life of 22 years and starts to decay after someone has died. Stuart Black of Reading University in the United Kingdom used this isotope to date a body that was found floating down London's River Thames in September 2001.

© Horacio Villalobos/CORBIS

a. Find an equation for a model for the percent of a lead-210 sample left after t years.

b. Estimate the percent of lead-210 left after 60 years.

c. Estimate graphically the number of years after which only 5% of the sample will be left.

22. The U.S. Surgeon General estimates that radon is responsible for nearly 30,000 lung cancer deaths in the United States each year. Radon-222 is a gas that occurs naturally as uranium-238 decays in the soil. Because radon-222 is a gas, it gets released into the air and can enter homes through cracks in their foundations or basements. This gas is colorless and odorless, so it cannot be detected easily by humans. Radon detectors (shown below) have now made it easier for homeowners to keep their homes safe from extreme radon levels. Radon-222 has a half-life of 3.825 days.

© Tony Freeman/Photo Edit

a. Find an equation for a model for the percent of a radon-222 sample left after d days.

b. Estimate the percent of radon-222 left after 30 days.

c. Estimate graphically the number of days after which only 5% of the radon-222 sample will be left.

23. A sample of 300 polonium-218 atoms is being stored for an experiment that will take place in 2 hours. Polonium-218 has a half-life of 3.05 minutes.

a. Find an equation for a model for the number of polonium-218 atoms left after m minutes.

b. Find the number of polonium-218 atoms remaining at the beginning of the experiment.

c. Estimate numerically when there were 100 polonium-218 atoms left.

24. Another radioisotope that forensic pathologists use to date skeletal remains is polonium-210. This element has a half life of only 138 days and can be used to fine-tune a date of death to within a two-week period.

a. Find an equation for a model for the percent of polonium-210 atoms left d days after the death of a person.

b. Estimate the percent of polonium-210 left 1 year after the death.

c. Estimate numerically the time of death for a skeleton that has 10% of the polonium-210 expected in a body at the time of death.

For Exercises 25 through 28, use the graphs to answer the questions.

25. Given the graph of $f(x)$, estimate the following.

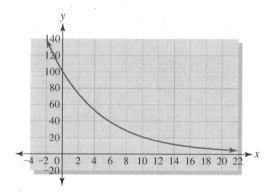

a. Is the graph increasing or decreasing?

b. Is this an example of exponential growth or exponential decay?

c. $f(10) = $.

d. What x value(s) will make $f(x) = 40$.

e. Estimate the y-intercept.

26. Given the graph of $h(x)$, estimate the following.

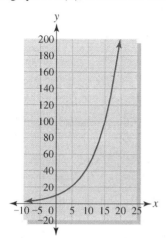

a. Is the graph increasing or decreasing?

b. Is this an example of exponential growth or exponential decay?

c. $h(5) = $.

d. What x value(s) will make $h(x) = 100$.

e. Estimate the y-intercept.

27. Given the graph of $f(x)$, answer the following.

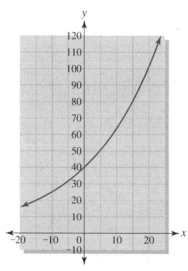

a. Is the graph increasing or decreasing?

b. Is this an example of exponential growth or exponential decay?

c. $f(-15) = $.

d. What x value(s) will make $f(x) = 80$.

e. Estimate the vertical intercept.

28. Given the graph of $g(x)$, find the following.

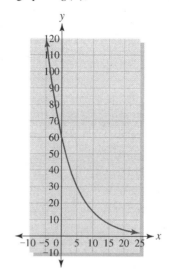

a. Is the graph increasing or decreasing?

b. Is this an example of exponential growth or exponential decay?

c. $g(7.5) = $.

d. What x value(s) will make $g(x) = 30$.

e. Estimate the vertical intercept.

For Exercises 29 through 50, find exponential models for the given data.

29.

x	f(x)
0	25
1	100
2	400
3	1600
4	6400

30.

x	f(x)
0	12
1	36
2	108
3	324
4	972

31.

x	f(x)
0	−35
1	−245
2	−1715
3	−12,005
4	−84,035

32.

x	f(x)
0	−20
1	−100
2	−500
3	−2500
4	−12,500

33.

x	f(x)
0	2000
1	400
2	80
3	16
4	3.2

34.

x	f(x)
0	360
1	180
2	90
3	45
4	22.5

35.

x	f(x)
0	32
1	48
2	72
3	108
4	162

36.

x	f(x)
0	81
1	135
2	225
3	375
4	625

37.

x	f(x)
0	6400
1	800
2	100
3	12.5
4	1.5625

38.

x	f(x)
0	2400
1	1800
2	1350
3	1012.5
4	759.375

39.

x	f(x)
0	3
5	12
10	48
15	192
20	768

40.

x	f(x)
0	7
2	42
4	252
6	1512
8	9072

41.

x	f(x)
0	−7
10	−21
20	−63
30	−189
40	−567

46.

x	f(x)
2	11
3	77
4	539
5	3773
6	26,411

42.

x	f(x)
0	−8
15	−32
30	−128
45	−512
60	−2048

47.

x	f(x)
5	2
6	12
7	72
8	432
9	2592

43.

x	f(x)
0	1701
6	567
12	189
18	63
24	21

48.

x	f(x)
7	−5
8	−30
9	−180
10	−1080
11	−6480

44.

x	f(x)
0	4352
3	3264
6	2448
9	1836
12	1377

49.

x	f(x)
4	3584
5	896
6	224
7	56
8	14

45.

x	f(x)
2	80
3	320
4	1280
5	5120
6	20,480

50.

x	f(x)
3	−7776
4	−1296
5	−216
6	−36
7	−6

Solving Equations Using Exponent Rules

RECAP OF THE RULES FOR EXPONENTS

In section 3.1 we went over the rules for exponents. In this section we will use some of these rules to solve different types of equations.

RULES FOR EXPONENTS

1. $x^m \cdot x^n = x^{m+n}$

2. $(xy)^m = x^m y^m$

3. $\dfrac{x^m}{x^n} = x^{m-n}$

4. $\left(\dfrac{x}{y}\right)^m = \dfrac{x^m}{y^m}$

5. $(x^m)^n = x^{mn}$

6. $x^0 = 1$

7. $x^{-n} = \dfrac{1}{x^n}$

8. $x^{1/n} = \sqrt[n]{x}$

SOLVING SIMPLE EXPONENTIAL EQUATIONS

Some simple exponential equations can be solved by inspection or trial and error. One property of exponents that is very helpful is that if two exponential expressions have the same base and are equal to one another, then the exponents must also be equal.

$$3^x = 9 \qquad \text{If possible, make the bases the same.}$$

$$3^x = 3^2$$

$$x = 2 \qquad \text{Because the bases are the same, the exponents themselves must be equal.}$$

We can use this basic property of exponents to solve simple exponential equations. It is important to remember that when you are solving an equation you want to recognize what type of equation you are working with. Exponential equations will always have a variable in the exponent. In this section we will learn to solve simple exponential equations using inspection or trial and error.

EXAMPLE **1** SOLVING SIMPLE EXPONENTIAL EQUATIONS

Solve the following exponential problems by inspection or trial and error.

a. $2^x = 16$

b. $3^x = 243$

c. $\dfrac{1}{1000} = 10^t$

d. $2^x = \dfrac{1}{8}$

Solution

a. $2^x = 16$ Write 16 as a power of 2.

$2^x = 2^4$ Because the bases are the same, the exponents must be equal.

$x = 4$

$2^4 = 16$ Check your solution.

$16 = 16$

b. Many people do not know what power of 3 will give us 243. In this case we will use trial and error and calculate different powers of 3 until we find the power that gives us 243.

$3^x = 243$ Write 243 as a power of 3.

$3^x = 3^5$ Because the bases are the same,
the exponents must be equal.

$x = 5$

$3^5 = 243$ Check your solution.

$243 = 243$

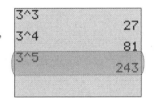

```
3^3
                    27
3^4
                    81
3^5
                   243
```

c. This exponential equation has a base of 10 but the other side of the equation is a fraction. This means that a negative exponent will be needed to get the reciprocal of the fraction.

$\dfrac{1}{1000} = 10^t$ Write both sides using the same base.
The bases on both sides of the equal sign must be the same and be in the same location. In this problem we will put both bases in the numerator.

$\dfrac{1}{10^3} = 10^t$

$10^{-3} = 10^t$ You will need negative exponents to get the reciprocal of the fraction so that the 10 cubed will come up to the numerator.

$-3 = t$

$\dfrac{1}{1000} = 10^{-3}$ Check your solution.

$\dfrac{1}{1000} = \dfrac{1}{1000}$

d. Again this exponential equation has a fraction, so a negative exponent will be needed to get a reciprocal of the fraction and thus make the bases the same.

$2^x = \dfrac{1}{8}$ Write both sides using the same base.

$2^x = \dfrac{1}{2^3}$

$2^x = 2^{-3}$ You will need negative exponents to get the reciprocal.

$x = -3$

    Check your solution.

$2^{-3} = \dfrac{1}{8}$

$\dfrac{1}{8} = \dfrac{1}{8}$

EXAMPLE PRACTICE PROBLEM

Solve the following exponential problems by inspection or trial and error.

a. $10^x = 100000$ **b.** $4^t = 4096$ **c.** $2^x = \dfrac{1}{4}$

Some exponential equations have more than the base and exponent on one side of the equation. When this happens we must first isolate the exponential part and then we can try to solve using inspection or trial and error.

EXAMPLE 2 SOLVING EXPONENTIAL EQUATIONS

Solve the following exponential problems by inspection or trial and error.

a. $5(3^t) = 45$ **b.** $6(4^x) - 34 = 350$ **c.** $640(2^t) + 9 = 29$

Solution

a.

$$5(3^t) = 45$$

$$\frac{5(3^t)}{5} = \frac{45}{5} \qquad \text{First isolate the exponential part by dividing by 5.}$$

$$3^t = 9$$

$$3^t = 3^2 \qquad \text{Write both sides using the same base.}$$

$$t = 2 \qquad \text{Because the bases are the same, the exponents must be equal.}$$

$$5(3^2) = 45 \qquad \text{Check your solution.}$$

$$5(9) = 45$$

$$45 = 45$$

b.

$$6(4^x) - 34 = 350$$

$$\underline{\qquad\; +34 \quad\; +34\qquad} \qquad \begin{array}{l}\text{First isolate the exponential part by adding} \\ \text{34 to both sides and dividing both sides by 6.}\end{array}$$

$$6(4^x) = 384$$

$$\frac{6(4^x)}{6} = \frac{384}{6} \qquad \text{Check your solution using the calculator.}$$

$$4^x = 64$$

$$4^x = 4^4$$

$$x = 4$$

Plot1 Plot2 Plot3
\Y1■6(4^X)-34
\Y2■350
\Y3=■
\Y4=
\Y5=
\Y6=
\Y7=

X	Y1	Y2
3	350	350
X=		

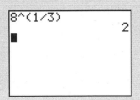
c.

$$640(2^t) + 9 = 29$$
$$\underline{\qquad -9 \quad -9 \qquad}$$
$$640(2^t) = 20$$

First isolate the exponential part by subtracting 9 from both sides and dividing both sides by 640.

$$\frac{640(2^t)}{640} = \frac{20}{640}$$

Reduce the fraction since the decimal form may not be easy to recognize as a power of 2.

$$2^t = \frac{1}{32}$$

Check your solution using the calculator.

$$2^t = \frac{1}{2^5}$$

$$2^t = 2^{-5}$$

$$t = -5$$

EXAMPLE ② PRACTICE PROBLEM

Solve the following exponential problems by inspection or trial and error.

a. $-2(5^x) - 74 = -324$

b. $162(3^x) + 30 = 36$

SOLVING POWER EQUATIONS

We will use rational exponents to solve equations that have variables raised to different powers. Raising both sides of an equation to the reciprocal exponent will help us to eliminate exponents of variables we are trying to solve for. Remember that a fraction exponent is another way of writing a radical so we are actually taking the appropriate root of both sides. The square root property that we learned in chapter 4 was an example of this process. Recall that the square root property required you to use a plus/minus symbol to account for both possible solutions. The plus/minus symbol is necessary whenever we take an even root or raise both sides to a fractional exponent where the denominator is an even number.

Odd power.	Even power requires plus/minus symbol.
$x^5 = 32$	$x^4 = 81$
$(x^5)^{1/5} = (32)^{1/5}$	$(x^4)^{1/4} = \pm(81)^{1/4}$
$x = 2$	$x = \pm 3$

EXAMPLE 3 SOLVING POWER EQUATIONS

Solve the following equations.

a. $78125 = x^7$

b. $150m^4 = 22509.375$

c. $\dfrac{t^6}{500} = 20$

d. $3t^5 - 17 = 3055$

Solution

a.

$$78125 = x^7$$ You need to undo an exponent of 7.

$$(78125)^{1/7} = (x^7)^{1/7}$$ Raise both sides of the equation to the reciprocal of 7.

$$5 = x$$

$$78125 = 5^7$$ Check your solution.

$$78125 = 78125$$

b.

$$150m^4 = 22509.375$$ Isolate the variable term by dividing by 150.

$$\frac{150m^4}{150} = \frac{22509.375}{150}$$

$$m^4 = 150.0625$$

$$(m^4)^{1/4} = \pm(150.0625)^{1/4}$$ Raise both sides to the reciprocal power. Because the exponent is an even power we need to use a plus/minus symbol.

$$m = \pm 3.5$$

$$150(3.5)^4 = 22509.375$$ Check your solutions.

$$22509.375 = 22509.375$$

$$150(-3.5)^4 = 22509.375$$

$$22509.375 = 22509.375$$

c.

$$\frac{t^6}{500} = 20$$ Isolate the variable term by multiplying by 500.

$$500\left(\frac{t^6}{500}\right) = 500(20)$$

$$t^6 = 10000$$

$$(t^6)^{1/6} = \pm(10000)^{1/6}$$ Raise both sides to the reciprocal power. Because the exponent is an even power we need to use a plus/minus symbol.

$$t \approx \pm 4.642$$

Check your solutions using the calculator.

```
Plot1 Plot2 Plot3
\Y1◘(X^6)/500
\Y2◘20
\Y3=
\Y4=
\Y5=
\Y6=
\Y7=
```

X	Y1	Y2
4.642	20.011	20
-4.642	20.011	20

X=

The Y1 and Y2 values are not exactly the same because of rounding.

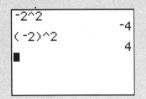

d.

$$-3t^5 - 17 = 3055 \qquad \text{Isolate the variable term.}$$
$$-3t^5 = 3072$$
$$t^5 = -1024$$
$$(t^5)^{1/5} = (-1024)^{1/5} \qquad \text{Raise both sides to the reciprocal power.}$$
$$t = -4$$

Check your solution on the calculator.

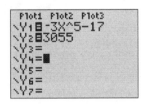

EXAMPLE ③ PRACTICE PROBLEM

Solve the following equations.

a. $-6x^9 + 148500 = 30402$ 　　　 **b.** $\dfrac{p^4}{30} = 7$

c. $2m^5 - 61 = 847.70848$

Remember that when you are solving equations to consider the type of equation you are working with. We use reciprocal powers to solve equations that have numbers as the exponents. If the variable is in the exponent the equation is an exponential and will not be solved using reciprocal exponents. In the next chapter we will learn a way to solve more exponential equations that are not as easy to solve using inspection or trial and error.

EXAMPLE 4 IDENTIFYING EXPONENTIAL AND POWER EQUATIONS

Determine if the following equations are exponential or power equations.

a. $3^x + 5 = 32$ 　　　 **b.** $x^7 + 616 = 17000$ 　　　 **c.** $4(5^x) - 23 = 1435$

Solution

a. Since the variable x is in the exponent, this is an exponential equation.

b. Since the variable x is being raised to a power, this is a power equation.

c. Again the variable x is in the exponent, so this is an exponential equation.

5.2 Exercises

For Exercises 1 through 10, determine if the given equation is a power equation or an exponential equation. Do not solve these equations.

1. $12^x = 45$

2. $7^x = 265$

3. $x^5 = 436$

4. $m^8 = 0.2586$

5. $3x^7 + 25 = 6986$

6. $154 + 36p^5 = 0.568$

7. $4(6^x) + 8(6^x) = 4586$

8. $2.3\left(\dfrac{1}{3}\right)^x - 4.8\left(\dfrac{1}{3}\right)^x - 96.8 = 42.3658$

9. $4.5m^{2/3} + 86 = 3316$

10. $2.8b^{3/4} = 2535 - 1.7b^{3/4}$

For Exercises 11 through 54, solve the following exponential equations using inspection or trial and error.

11. $2^x = 8$

12. $4^x = 16$

13. $5^c = 625$

14. $10^t = 10{,}000$

15. $\dfrac{1}{9} = 3^t$

16. $5^x = \dfrac{1}{25}$

17. $3^x = \dfrac{1}{81}$

18. $2^r = \dfrac{1}{8}$

19. $(-2)^d = -32$

20. $(-5)^x = -125$

21. $(-6)^w = 36$

22. $(-7)^x = 49$

23. $\left(\dfrac{1}{2}\right)^x = \dfrac{1}{16}$

24. $\left(\dfrac{1}{3}\right)^t = \dfrac{1}{9}$

25. $\left(\dfrac{1}{2}\right)^x = 16$

26. $\left(\dfrac{1}{3}\right)^t = 9$

27. $\left(\dfrac{1}{2}\right)^x = 32$

28. $\left(\dfrac{1}{5}\right)^t = 25$

29. $10^x = 1000$

30. $10^c = 100{,}000{,}000$

31. $10^x = 1$

32. $3^x = 1$

33. $(-5)^m = 1$

34. $(-7)^x = 1$

35. $\left(\dfrac{2}{3}\right)^t = 1$

36. $\left(\dfrac{4}{7}\right)^x = 1$

37. $5(2^x) = 40$

38. $7(3^x) = 189$

39. $-2(5^c) = -250$

40. $-4(3^t) = -36$

41. $3^c + 5 = 32$

42. $2^x + 30 = 62$

43. $7^t + 8 = 57$

44. $5^m - 200 = 425$

45. $-4(5^m) - 9 = -109$

46. $-6(2^t) - 25 = -409$

47. $3(6^x) + 2(6^x) = 180$

48. $4(5^k) + 8(5^k) = 1500$

49. $10(2^x) - 7(2^x) = 48$

50. $5(3^x) - 4(3^x) = 81$

51. $7(2^x) = 48 + 4(2^x)$

52. $12(6^t) = -252 + 19(6^t)$

53. $13(7^w) + 20 = 5(7^w) + 2764$

54. $4(3^k) - 40 = -6(3^k) + 230$

For Exercises 55 through 72, solve the following power equations.

55. $x^5 = 32$

56. $t^3 = 64$

57. $x^6 = 15625$

58. $p^4 = 2401$

59. $12w^4 = 7500$

60. $7t^8 = 45927$

61. $-7x^3 = 1512$

62. $-2.5m^7 = 320$

63. $3x^5 + 94 = 190$

64. $4m^7 + 20 = 380$

65. $5x^6 - 30 = 1475$

66. $6t^8 + 18 = 393234$

67. $96 - 24c^4 = 40$

68. $74 - 3x^{12} = -2457$

69. $\dfrac{4x^3 + 5}{8} = 63.125$

70. $\dfrac{2m^5 + 38}{16} = 32.75$

71. $\dfrac{7b^4 + 58}{8} = 1141.25$

72. $\dfrac{2.4x^6 - 19}{1000} = 282.3386$

For Exercises 73 through 84, determine if the equation is a power equation or an exponential equation and solve.

73. $3^k = 81$

74. $6^b = 216$

75. $150r^5 = 1725$

76. $13p^7 = 101.4$

77. $3(4^x) + 20 = 212$

78. $8(3^x) + 74 = 722$

79. $45 - 3.5g^5 = -67$

80. $14 - 7.8x^5 = 1836.5$

81. $\dfrac{3.6h^8 - 56}{33} = 58.168$

82. $\dfrac{23.5h^4 - 75}{56} = 61.63$

83. $\left(\dfrac{1}{5}\right)^x = \dfrac{1}{125}$

84. $\left(\dfrac{1}{4}\right)^t = \dfrac{1}{16}$

Graphing Exponential Functions

In this section we will use the concept of an exponential function, the rules for exponents and the solving techniques that we reviewed in Section 5.2 to develop a method for finding an exponential model. It is crucial that you recognize the graph of an exponential model because we always start with a scatterplot of data and choose a model from the shape the data takes. With this in mind, let's look at a few exponential graphs and describe their basic characteristics.

Recall from Section 5.1 that a basic exponential function has the form

$$f(x) = a \cdot b^x$$

where a, and b are real numbers, $a \neq 0$, $b > 0$, and $b \neq 1$.

EXPLORING GRAPHS OF EXPONENTIALS

 CONCEPT INVESTIGATION I WHAT DO a AND b DO TO THE GRAPH?·······················

Consider the standard form of an exponential function $f(x) = a \cdot b^x$. We are going to investigate one part of this function at a time. For each section of this investigation, consider how the graph changes as you change one of the constants in the function.

a. Graph the following exponential functions on the same calculator window. (Use the window Xmin $= -7$, Xmax $= 7$, Ymin $= -10$, and Ymax $= 100$. Remember to turn STATPLOT off.)

   i. $f(x) = 2^x$

   ii. $f(x) = 5^x$

   iii. $f(x) = 10^x$

   iv. $f(x) = 22.5^x$

Describe the basic shape that all of these graphs have.

In these examples we are considering the shape of the graph of a basic exponential function when the value of the base b is greater than 1. In your own words, describe how increasing the value of b affects the graph.

b. Graph the following exponential functions on the same calculator window. (Use the window Xmin $= -7$, Xmax $= 7$, Ymin $= -10$, and Ymax $= 100$.)

   i. $f(x) = \left(\dfrac{1}{2}\right)^x$

   ii. $f(x) = \left(\dfrac{1}{5}\right)^x$

   iii. $f(x) = (0.1)^x$

   iv. $f(x) = (0.025)^x$

In these functions we are considering how having $0 < b < 1$ changes the graph of a basic exponential function. Recall that the base of an exponential function must be positive and cannot equal 1. In your own words, what does a value of b less than 1 do to the graph?

c. Graph the following exponential functions on the same calculator window. (Use the window Xmin $= -7$, Xmax $= 7$, Ymin $= -10$, and Ymax $= 100$.)

  i. $f(x) = (2)^x$

  ii. $f(x) = 5(2)^x$

  iii. $f(x) = 10(2)^x$

  iv. $f(x) = 8\left(\dfrac{1}{2}\right)^x$

In these functions we are considering how a positive a value changes the graph of a basic exponential function. In your own words, what does a positive a value do to the graph?

d. Graph the following exponential functions on the same calculator window. (Use the window Xmin $= -7$, Xmax $= 7$, Ymin $= -100$, and Ymax $= 10$.)

  i. $f(x) = -1(2)^x$

  ii. $f(x) = -5(2)^x$

  iii. $f(x) = -10(2)^x$

  iv. $f(x) = -8\left(\dfrac{1}{2}\right)^x$

In these functions we are considering how a negative a value changes the graph of a basic exponential function. In your own words, what does a negative a value do to the graph?

◆

Using all of the information from this concept investigation should allow us to sketch a graph of a basic exponential function and adjust any models that we are going to make. You should note that the graph of an exponential function of the form $f(x) = ab^x$ never crosses the horizontal axis. The graph does get increasingly close to the horizontal axis but does not touch it. This is because an exponential function of this form can never equal zero. This makes the horizontal axis a **horizontal asymptote** for this graph. A horizontal asymptote is any horizontal line that a graph gets increasingly close to but does not touch. Other functions that we will study may have horizontal asymptotes along with vertical asymptotes.

GRAPHS OF EXPONENTIAL FUNCTIONS

Positive a values.

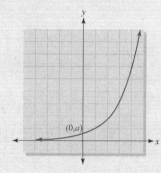

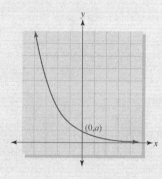

$$f(x) = a \cdot b^x$$
$$a > 0$$
$$b > 1$$

$$f(x) = a \cdot b^x$$
$$a > 0$$
$$0 < b < 1$$

Negative a values.

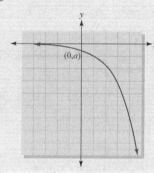

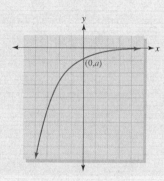

$$f(x) = a \cdot b^x$$
$$a < 0$$
$$b > 1$$

$$f(x) = a \cdot b^x$$
$$a < 0$$
$$0 < b < 1$$

EXAMPLE 1 GRAPHING EXPONENTIAL FUNCTIONS WITH POSITIVE a VALUES.

Sketch the graph of the following functions, by hand. Discuss any information you know from the values of a and b.

a. $f(x) = 5(3)^x$

b. $g(x) = 255\left(\dfrac{2}{3}\right)^x$

Solution

a. $a = 5$, so the vertical intercept is $(0, 5)$, and the graph will be above the horizontal axis. The base $b = 3$, so this graph is increasing, showing exponential growth.

By trying a few input values, we get a table of points. Using these points to plot the graph, we get the following graph

x	f(x)
1	15
2	45
3	135

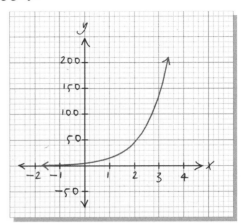

b. $a = 255$, so the vertical intercept is $(0, 255)$, and the graph will be above the horizontal axis. The base $b = \frac{2}{3}$, so this graph is decreasing, showing exponential decay. By trying a few input values, we get a table of points. Using these points, we get the following graph.

x	g(x)
−2	573.75
1	170
2	113.33

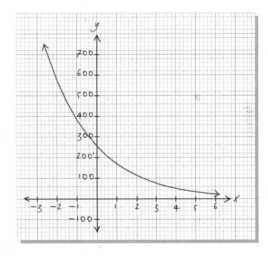

EXAMPLE PRACTICE PROBLEM

Sketch the graph of the following functions, by hand. Discuss any information you know from the values of a and b.

a. $f(x) = 8(1.25)^x$

b. $g(x) = 300(0.5)^x$

From the concept investigation above we saw that when a is a negative number the graph is below the horizontal axis. The graph has actually been reflected (flipped) over the horizontal axis. The graph is a mirror image of the same graph with the opposite *a* value. This reflection makes graphing an exponential function with a negative *a* value easier. If *a* is negative you can imagine the graph as if *a* were actually positive and then flip the graph over the horizontal axis. If we consider the function $f(x) = -10(2)^x$ we see that $a = -10$ so the graph will be below the horizontal axis. To start graphing, we could pretend that *a* is positive and sketch the graph. Since $b = 2$ it is greater than 1, so the graph would be exponential growth. Using a positive *a* value and the fact that this is exponential growth gives us the blue dashed curve. Flipping the blue dashed curve over the horizontal axis gives us the final graph of $f(x) = -10(2)^x$.

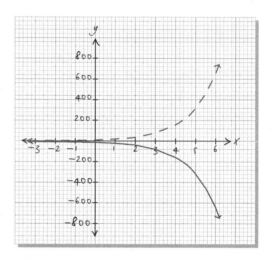

EXAMPLE 2 GRAPHING EXPONENTIAL FUNCTIONS WITH NEGATIVE a VALUES

Sketch the graph of the following functions, by hand. Discuss any information you know from the values of *a* and *b*.

a. $h(x) = -3(5)^x$

b. $W(x) = -50(0.2)^x$

Solution

a. $a = -3$, so the vertical intercept is $(0, -3)$, and the graph will be below the horizontal axis. The base $b = 5$, so this graph should show exponential growth, but it has been flipped under the horizontal axis by the negative *a* value. By trying a few input values, we get a table of points. Using these points we get the following graph.

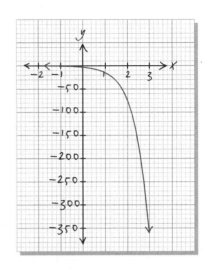

x	h(x)
−1	−0.6
1	−15
2	−75

b. $a = -50$, so the vertical intercept is $(0, -50)$, and the graph will be below the horizontal axis. The base $b = 0.2$, so this graph should show exponential decay, but it has also been flipped under the horizontal axis by the negative a value. By trying a few input values, we get a table of points. Using these points we get the following graph.

x	W(x)
−1	−250
1	−10
2	−2

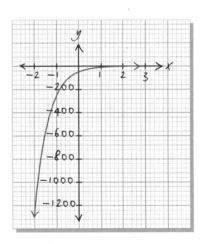

EXAMPLE   ②   PRACTICE PROBLEM

Sketch the graph of the following functions, by hand. Discuss any information you know from the values of a and b.

a. $f(x) = -2(2.5)^x$

b. $h(x) = -145(0.5)^x$

DOMAIN AND RANGE OF EXPONENTIAL FUNCTIONS

Exponential functions, like the linear and quadratic functions that we have already studied, have a pretty standard domain and range when the domain is not restricted by a real world context. Consider a basic graph of an exponential function

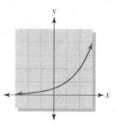

In this graph, you should notice that the function is valid for all input values but does not hit all outputs. This exponential growth function does not have any negative output values. A basic exponential function cannot equal zero, so that cannot be part of the range for the function. The curve will get as close to the horizontal axis as possible but will never touch it. For this graph the horizontal axis is called a horizontal asymptote. Because this exponential function only has positive output values, the range will be only the positive real numbers.

In general, the domain of an exponential function that has no context will be all real numbers, or $(-\infty, \infty)$. The range for an exponential function will either be all positive or all negative real numbers, $(-\infty, 0)$ or $(0, \infty)$. The range will be positive whenever the value of a is positive and negative when the value of a is negative. Making a quick sketch of the graph of the function can help you determine the correct domain and range.

When a is positive.

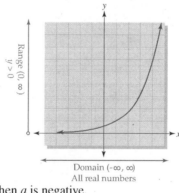

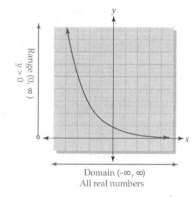

When a is negative.

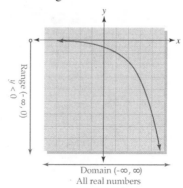

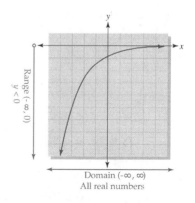

Remember that the domain and range in a context should take into consideration possible model breakdown in the situation. Model breakdown will limit the domain and range for the models that we find in later sections.

EXAMPLE 3 DOMAIN AND RANGE OF EXPONENTIAL FUNCTIONS

Find the domain and range of the following exponential functions.

a. $f(x) = 5(1.25)^x$

b. $g(x) = -30(0.7)^x$

c. $h(x) = 400(0.8)^x$

Solution

a. This function will be exponential growth because the base is larger than 1. It will also be positive since a is positive. A quick sketch of the graph would give us

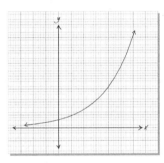

The sketch of the graph confirms the following domain and range.

Domain: All real numbers or $(-\infty, \infty)$

Range: All positive real numbers or $(0, \infty)$.

b. This function would be exponential decay because the base is less than 1 but a is negative so the graph will be flipped below the horizontal axis. A quick sketch of the graph would give us

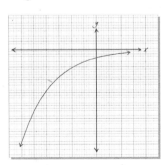

The sketch of the graph confirms the following domain and range.

Domain: All real numbers or $(-\infty, \infty)$

Range: All negative real numbers or $(-\infty, 0)$.

c.   This function will be exponential decay since the base is less than 1 and positive since *a* is positive. With this information we know the function has the following domain and range.

Domain: All real numbers or $(-\infty, \infty)$

Range: All positive real numbers or $(0, \infty)$.

EXAMPLE ③ PRACTICE PROBLEM

Find the domain and range of the following exponential functions.

a. $f(x) = -3(2.1)^x$

b. $h(x) = 250(0.5)^x$

5.3 Exercises

For Exercises 1 through 10, use the graphs of the exponential functions $f(x) = ab^x$ to give the following information.

   a. *Is a positive or negative? Explain.*

   b. *Is b greater than or less than 1? Explain.*

   c. *Domain of the function.*

   d. *Range of the function.*

1.

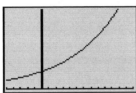

2.

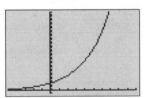

3.

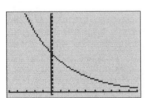

4.

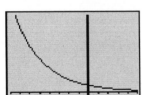

5.

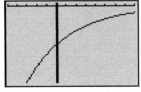

6.

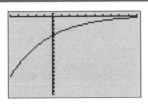

7.

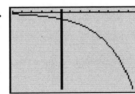

8.

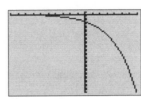

9.

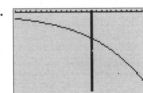

10.

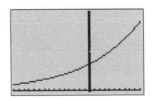

For Exercises 11 through 32, sketch the graph of the functions by hand, state any information you know from the values of a *and* b, *and give the domain and range of the function.*

11. $f(x) = 7(2)^x$

12. $h(x) = 5(3)^x$

13. $g(x) = 3(1.2)^x$

14. $r(x) = 4(1.5)^x$

15. $h(x) = 12(1.4)^x$

16. $f(x) = 20(1.15)^x$

17. $p(t) = 140\left(\dfrac{1}{2}\right)^t$

18. $r(t) = 100\left(\dfrac{2}{5}\right)^t$

19. $f(x) = 250\left(\dfrac{1}{4}\right)^x$

20. $g(x) = 400\left(\dfrac{3}{4}\right)^x$

21. $f(x) = -2(1.4)^x$

22. $h(x) = -6(1.25)x$

23. $g(t) = -3(0.7)^t$

24. $f(x) = -8(0.6)^x$

25. $h(m) = 0.5(2.5)^m$

26. $m(t) = 0.75(4)^t$

27. $j(w) = -0.5(4)^w$

28. $g(x) = -10(2)^x$

29. $h(t) = -0.4(1.5)^t$

30. $m(t) = -0.5(1.75)^t$

31. $c(n) = 550(0.75)^n$

32. $f(t) = 200\left(\dfrac{3}{5}\right)^t$

Finding Exponential Models

FINDING EXPONENTIAL MODELS

The steps to model exponential data are similar to the other modeling processes we have done. The exponential form $f(x) = a \cdot b^x$ requires values for a and b. This requirement is similar to the linear model when we found m and b, or the quadratic model when we found a, h and k. For exponential modeling we will use a system of equations and a version of the elimination method to solve for b. This process will require us to use the solving techniques we learned in section 5.2. We will use the following steps to model data that have exponential characteristics.

MODELING STEPS FOR EXPONENTIALS
(WITH NO GIVEN VERTICAL INTERCEPT)

1. Define the variables and adjust the data.
2. Create a scatterplot.
3. Select a model type.
4. **Exponential Model:** Pick two points, write two equations using the form $f(x) = a \cdot b^x$ and solve for b.
5. Use b and one equation to find a.
6. Write the equation of the model using function notation.
7. Check your model by graphing it with the scatterplot.

EXAMPLE 1 FINDING AN EXPONENTIAL MODEL

Software development is undergoing a major change from a closed software development process to a process that uses open source software. The total numbers of source lines of code in millions are given in the following bar chart.

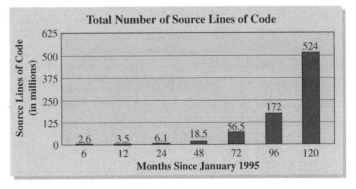

Source: Data estimated from information found at www.riehle.org.

a. Find an equation for a model for these data.

b. Estimate the total number of source lines of code 12 years after January 1995.

Solution

a.

Step 1 Define the variables and adjust the data.

m = Months since January 1995.

L = Total number of source lines of code in millions.

These data do not have to be adjusted.

Step 2 Create a scatterplot.

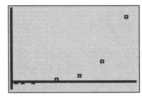 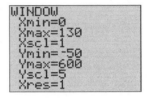

Step 3 Select a model type.

As you can see, the graph is curved up much like a quadratic but seems to be flatter on the left-hand side of the data than a quadratic should be. A quadratic model would go up to the left of a vertex and therefore would not be a good choice here. Because of this shape and type of growth, we should choose an exponential model.

Step 4 **Exponential Model:** Pick two points, write two equations using the form $f(x) = a \cdot b^x$ and solve for b.

We can pick the first and last points of these data because they appear to fall along the exponential path and one is in the flat section of the curve while the other is in the steeper part.

$$(6, 2.6) \quad \text{and} \quad (120, 524)$$

<table>
<tr><td>$2.6 = a \cdot b^6 \qquad 524 = a \cdot b^{120}$</td><td>Use the two points to write two equations in standard form. We now have a system of two equations with the variables multiplied together.</td></tr>
</table>

$$\frac{524}{2.6} = \frac{a \cdot b^{120}}{(a \cdot b)^6}$$

Therefore, we divide the two equations and solve for b.

$$201.54 \approx b^{114}$$

$$(201.54)^{1/114} \approx (b^{114})^{1/114}$$

Use the reciprocal exponent to isolate b.

$$1.048 \approx b$$

Step 5 Use b and one equation to find a.

$$2.6 = a(1.048)^6$$

Substitute b into one equation and solve for a.

$$2.6 \approx 1.325a$$

$$\frac{2.6}{1.325} \approx \frac{1.325a}{1.325}$$

$$1.96 \approx a$$

Step 6 Write the equation of the model using function notation.

$$L(m) = 1.96(1.048)^m$$

Step 7 Check your model by graphing it with the scatterplot.

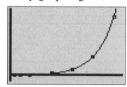

This model fits the data very well. It follows the flat pattern on the left side and then grows along with the data as you go to the right.

b. 12 years after January 1995 is represented by $m = 144$, so we can substitute 144 into the model and get

$$L(144) = 1.96(1.048)^{144}$$
$$L(144) = 1676$$

Therefore the 12 years after January 1995 there were a total of 1,676 million source lines of code.

EXAMPLE ① PRACTICE PROBLEM

Find an equation for a model for the data given in the table.

x	f(x)
2	4.8
3	7.0
5	14.7
6	21.4
8	44.9
10	94.5
12	198.7

DOMAIN AND RANGE FOR EXPONENTIAL MODELS

In working with exponential models in real-world applications, the domain and range will again depend on the situation and require us to avoid model breakdown. Because exponential functions grow or decay very quickly, the domain will usually be very restricted to values close to the original data. In a context, always try to expand the domain beyond the data but be cautious of output values that get too extreme for the context of the problem.

The range for an exponential model must come from the domain, and because the endpoints of the graph will always be its lowest and highest values, the range should always come from the endpoints of the domain just as they did for lines. Remember to always write the domain and range as lowest value to highest value.

EXAMPLE DOMAIN AND RANGE OF AN EXPONENTIAL MODEL

What would a reasonable domain and range be for the model found in example 1?

Solution

In this case the total number of source lines of code would be low before January 1995 and continue to grow for some time after 2005, so a reasonable domain would probably be [3, 144]. In this case the lowest point is on the left endpoint of $m = 3$, and the highest point is on the right endpoint at $m = 144$. Substituting in these values for m gives us a range of [2.26, 1676]. Because the curve shows fairly fast growth, it will most likely not be able to sustain that kind of growth for long. Be careful when picking domain so that you do not extend too far in either direction which may result in model breakdown.

When finding an exponential model, having a vertical intercept to take from the data can simplify the modeling steps. When you have a vertical intercept to take from the data, you will not need two equations to solve for b. The output value of the vertical intercept will be the a value for the model and you can use another point and an equation to find b.

> **MODELING STEPS FOR EXPONENTIALS**
> **(WITH A GIVEN VERTICAL INTERCEPT)**
> 1. Define the variables and adjust the data (if needed).
> 2. Create a scatterplot.
> 3. Select a model type.
> 4. **Exponential Model:** Pick the vertical intercept and one other point.
> 5. Substitute the output value of the vertical intercept into a and use the other point to find b.
> 6. Write the equation of the model using function notation.
> 7. Check your model by graphing it with the scatterplot.

EXAMPLE 3 FINDING AN EXPONENTIAL MODEL GIVEN A VERTICAL INTERCEPT

The invasion of certain weeds can be a devastating problem for an agricultural area. The numbers of buckthorn plants found on a certain acre of land during a four-year period are given in the table.

Years	Number of Plants
0	2
0.5	6
1	16
1.5	45
2	126
2.5	355
3	1001

Concept Connection

Remember when reading data that the intercepts will occur when one of the variables equals zero.

The vertical intercept will occur whenever the input variable is zero.

The horizontal intercept(s) will occur whenever the output variable is zero.

In Example 3 we can see that the year zero is given so the vertical intercept for this model is (0, 2)

a. Find an equation for a model for these data.

b. Estimate the number of buckthorn plants after four years.

c. Give a reasonable domain and range for this model.

Solution

a.   **Step 1** Define the variables and adjust the data.

P = The number of buckthorn plants on this acre

t = The years since the start of the invasion

We do not need to adjust the data, since the data are already reasonable.

Step 2 Create a scatterplot.

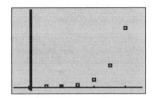   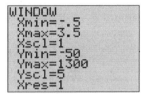

Step 3 Select a model type.
This graph is again flat on the left side and rises quickly to the right, so it seems to be exponential.

Step 4 Exponential Model: Pick the vertical intercept and one other point.
The vertical intercept is given as (0, 2), and another point could be the last one at (3, 1001). If you choose two points that are too close together, you might not get a reasonable value for b. In this case, one of the last few points in the data should work well.

Step 5 Substitute the output value of the vertical intercept into a and use the other point to find b.

$$P = 2 \cdot b^t$$ Substitute the vertical intercept into a.
$$1001 = 2b^3$$ Substitute the other point in for P and t.
$$500.5 = b^3$$ Isolate the variable term.
$$(500.5)^{1/3} = (b^3)^{1/3}$$ Solve for b using the reciprocal exponent.
$$7.94 \approx b$$

Step 6 Write the equation of the model using function notation.

$$P(t) = 2(7.94)^t$$

Step 7 Check your model by graphing it with the scatterplot.

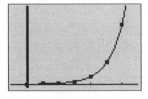

This seems to be a great fit for these data.

b. The fourth year can be represented as $t = 4$, so we get

$$P(4) = 2(7.94)^4$$
$$P(4) = 7949$$

This means that in year 4 there were approximately 7,949 buckthorn plants on the acre of land.

c. It makes sense to start our domain at $t = 0$, since that is the start of the invasion. Because the amounts of land and water available to the weeds are limited, we should not extend the domain much beyond the data. The answer for part *b* seems like a possible limit to this situation, so we will set the end of the domain to be $t = 4$. Thus for the domain of [0, 4] we get a range of [1, 7949].

Example 3 shows you the second option for finding an exponential model. This method is easier because you already have a value for *a* in the given vertical intercept, so you need to use only one equation to help you find *b*.

EXAMPLE ③ PRACTICE PROBLEM

Find an equation for a model for the data given in the table.

x	f(x)
0	145
3	74.2
5	47.5
8	24.3
10	14.6
14	6.4
15	5.1

5.4 Exercises

For Exercises 1 through 10, graph the data and the function on your calculator and adjust a and/or b to get an eye-ball best fit. Remember your answers will vary from the back of the book.

1. $f(x) = 5.5(1.2)^x$

x	f(x)
0	3.5
2	5
5	9
8	15
10	22
14	45

2. $f(x) = 6(1.6)^x$

x	f(x)
-2	3
0	7
3	29
5	73
7	188
11	1232

3. $f(x) = 100(0.45)^x$

x	f(x)
-1	173
0	78
1	35
3	7
4	3.2
6	0.65

4. $f(x) = 500(0.75)^x$

x	f(x)
-1	667
0	500
2	281
3	211
5	119
7	67

5. $f(x) = 8(1.5)^x$

x	f(x)
-2	5
0	8
1	10
3	18
5	30
8	65

6. $f(x) = 12(1.3)^x$

x	f(x)
-1	11
0	12
2	14.5
5	19
9	28
14	46

7. $f(x) = 120(0.4)^x$

x	f(x)
-1	240
0	120
1	60
3	15
4	7.5
6	2

8. $f(x) = 90(0.8)^x$

x	f(x)
-1	129
0	90
1	63
2	44
4	22
5	15

9. $f(x) = 6(1.4)^x$

x	f(x)
-1	4
0	5
2	7
3	8.5
5	12.5
10	31

10. $f(x) = 140(0.9)^x$

x	f(x)
-1	188
1	120
3	77
5	49
7	31
10	16

11. The growth of the Internet has outpaced anything most people had imagined. Data for the number of Internet hosts in the early 1990s is given in the chart.

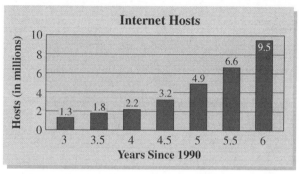

Source: *Matthew Gray, Massachusetts Institute of Technology.*

  a. Find an equation for a model of these data.

  b. Estimate the number of Internet hosts in 1997.

  c. Give a reasonable domain and range for this model.

  d. Estimate graphically when the number of Internet hosts reached 20 million.

12. In 1959 the United States started building intercontinental ballistic missiles (ICBMs) to act as a deterrent during the Cold War. The number of ICBMs in the U.S. arsenal is given in the chart.

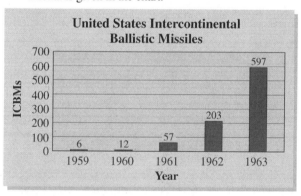

Source: *Natural Resources Defense Council.*

  a. Find an equation for a model of these data.

  b. Estimate the number of ICBMs in the U.S. arsenal in 1964.

  c. Give a reasonable domain and range for this model.

  d. Estimate numerically the year in which there would be 2500 ICBMs in the U.S. arsenal.

13. The number of nuclear warheads in the U.S. arsenal from 1945 to 1960 was growing at an exponential rate. The number of stockpiled warheads in the United States is given in the table.

Year	Stockpiled Warheads
1945	6
1947	32
1949	235
1951	640
1953	1436
1955	3057
1957	6444
1959	15,468

Source: Natural Resources Defense Council.

a. Find an equation for a model of these data.

b. Estimate the number of stockpiled warheads in 1950.

c. Give a reasonable domain and range for this model.

d. Estimate graphically the year in which the number of stockpiled warheads surpassed 50,000.

14. The number of weblogs has been growing very rapidly since early 2003. The estimated number of weblogs for years since January 2003 are given in the table.

Time (years)	Weblogs (millions)	Time (years)	Weblogs (millions)
0.5	.75	2.5	10.4
1	1.7	3	24
1.5	2.8	3.5	44
2	6	4	55

Source: Data estimated from information at www.sifry.com and Technorati

a. Find an equation for a model of these data.

b. Estimate the number of weblogs in January 2008.

c. Give a reasonable domain and range for this model.

d. Estimate graphically the year in which the number of weblogs surpassed 100 million.

15. While running an experiment in physics class, a student was testing the volume of water remaining in a cylinder after a small hole was made in the bottom. The data collected by the student are given in the table below.

Time (in seconds)	Volume (in liters)	Time (in seconds)	Volume (in liters)
30	13.24	180	5.32
60	11.45	210	4.34
90	9.96	240	3.59
120	7.79	270	3.20
150	6.10	300	2.66

a. Find an equation for a model of these data.

b. Estimate the water remaining after 6 minutes.

c. Give a reasonable domain and range for this model.

d. Estimate numerically how long it took for there to be only 1 liter of water remaining in the cylinder.

16. One concern all people should share is the overuse of our natural resources due to larger populations. The amount of organic material in our soils is crucial to the growth of food and other plants. With overuse, the percentage of soil that is organic material decreases to dangerous levels. Data demonstrating this loss after years of overuse are given in the table.

Year	Percent of Organic Material in Soil
0	3
25	2.1
75	0.9
125	0.4
175	0.2

Source: Data estimated from information from the University of Wisconsin.

a. Find an equation for a model of these data.

b. Estimate the percent of soil that is organic material after 100 years of overuse.

c. Give a reasonable domain and range for this model.

17. After a significant rainfall event the river gage level will typically decrease exponentially until it reaches its normal level. The following river gage levels were taken after such a rainfall event.

Days After Rainfall Event	River Gage Height (feet above normal)
0	14
1	8
2	4
3	2.4
4	1.25
5	.7
6	.4
7	0.25

a. Find an equation for a model of these data.

b. How far above normal will the gage height be 10 days after the rainfall event.

c. Give a reasonable domain and range for this model.

18. Students in a chemistry class were asked to run an experiment that would simulate the decay of a imaginary element. The students were given 100 dice to roll all at once and after each roll they would remove any die that showed a 5 on top. The students in one group collected the following data.

Number of Rolls	Remaining Dice
0	100
1	84
2	69
3	55
4	48
5	42
6	33
7	28

a. Find an equation for a model of these data.

b. How many dice will remain after 10 rolls?

For Exercises 19 through 34, Find an equation for a model for the given data and give the domain and range for the model.

19.

x	$f(x)$
0	4
2	7
5	16
7	28
11	85
13	148

20.

x	$f(x)$
0	7
3	38
4	67
9	1134
11	3514
13	10883

21.

x	$f(x)$
0	94
2	39
4	16
7	4
10	1
13	0.25

22.

x	$f(x)$
0	127
3	73
4	60
6	42
9	24
11	16

23.

x	f(x)
0	−17
1	−21
4	−38
6	−56
10	−124
13	−226

24.

x	f(x)
0	−23
5	−130
9	−412
15	−1307
25	−4142
40	−13123

25.

x	f(x)
0	−56
1	−25.2
3	−5.103
5	−1.033
7	−0.2093
9	−0.0424

26.

x	f(x)
0	−89
2	−46
5	−17.2
6	−12.4
8	−6.4
11	−2.4

27.

x	f(x)
5	14
7	23
12	76
15	155
18	318
20	512

28.

x	f(x)
2	7
5	90
7	477
9	2522
11	13339
12	30680

29.

x	f(x)
−1	855.56
3	73.75
7	6.19
10	0.97
11	0.53
13	0.15

30.

x	f(x)
−3	1112.20
−1	256.25
2	28.34
5	3.13
8	0.35
11	0.04

31.

x	f(x)
−5	−84.25
−3	−129.5
1	−306.3
4	−583.96
8	−1380.61
10	−2122.82

32.

x	f(x)
−6	−3.66
−2	−16.17
2	−71.49
5	−217.93
7	−458.20
11	−2025.48

33.

x	f(x)
−4	−1129.95
−1	−282.54
3	−44.51
5	−17.67
7	−7.01
9	−2.78

34.

x	f(x)
−2	−28291.02
−1	−9053.13
2	−296.65
4	−30.38
6	−3.11
9	−0.10

Exponential Growth and Decay Rates and Compounding Interest

EXPONENTIAL GROWTH AND DECAY RATES

Exponential growth and decay are often measured as a percentage change per year or other time period. Since a percentage is calculated using multiplication, if the same percentage growth or decay occurs over several years we will have repeated multiplication and thus an exponential pattern. The percentage change is often called a **growth rate**. Use the following concept investigation to compare two different growth patterns.

CONCEPT INVESTIGATION I WHAT PATTERN GROWS FASTER? •••••••••••• •••••••••••••••••

In this investigation we will consider two salary options for a companies employees.

Option 1: Starting salary of $50,000 with a $2000 raise each year.

Option 2: Starting salary of $50,000 with a 4% raise each year.

Complete the following table of calculations.

Years on the job	Option 1 Salary (dollars)	Option 2 Salary (dollars)
0	50,000	50,000
1	50,000+2,000=52,000	50,000+50,000(0.04)=52,000
2	52,000+2,000=54,000	52,000+52,000(0.04)=54,080
3		
4		
5		
6		

According to these calculations, which of the salary options is best for the employees?

Option 1 has a linear growth pattern because we are adding the same amount with each change in the number of years. Option 2 is not linear since we are not adding the same amount with each change in the years. This second option is actually exponential because we are multiplying by the same percentage each year.

Concept Connection

When calculating both of these options we need to remember to use the previous year's salary each time we do the calculation.

With option 2 we take the previous year's salary and add the 4% growth to it. We are calculating the 4% growth by multiplying the previous year's salary by 0.04. Because the 4% is based on a new amount each year the growth is larger each year.

Find an equation for a linear model for the option 1 salary.

Find an equation for an exponential model for the option 2 salary. (You can use the modeling process from section 5.4.)

Graph both salary option models on the same calculator screen and describe what happens over a 40 year career.

When you work with a percentage growth, the growth rate can be seen in the base of the exponential function. Consider the next example and the exponential model you found in the concept investigation above to see how the growth rate becomes a part of the base.

EXAMPLE 1 EXPONENTIAL GROWTH RATE IN NATURE

An ant colony can grow at a very rapid rate. One such colony studied by Deborah Gordon started with about 450 worker ants and grew 79% per year.

Source: Information taken from "The Development of Organization in an Ant Colony" by Deborah M. Gordon, American Scientist, *Jan–Feb 1995.*

a. Find an equation for a model for the number of worker ants in the colony.

b. Estimate the number of worker ants in the colony after five years.

c. In the sixth year of the study, the colony had about 10,000 worker ants. Does your model predict this many worker ants? Explain possible reasons why or why not.

Solution

a. To find an equation for a model, either we need two points to work with or we need to see a pattern to follow to build the model. We can start by building a small table of values from the information given. Because the colony grew by 79% each year, we can find the size of the colony by calculating the growth and adding that to the original population.

Let's start by defining variables.

A = The number of worker ants in the colony

t = Years after the study began

t	Calculation	$A(t)$
0		450
1	450 + 0.79(450)	805.5
2	805.5 + 0.79(805.5)	1441.8
3	1441.8 + 0.79(1441.8)	2580.9

The pattern here might be hard to see, so we will use two points and find the model. Since the first point (0, 450) is a vertical intercept, we will use it to simplify the process.

$$(0, 450) \quad \text{and} \quad (2, 1441.8)$$
$$450 = ab^0 \qquad 1441.8 = ab^2$$
$$a = 450 \qquad 1441.8 = ab^2$$
$$1441.8 = 450b^2$$
$$\frac{1441.8}{450} = \frac{450b^2}{450}$$
$$3.204 = b^2$$
$$\pm\sqrt{3.204} = b$$
$$\pm 1.79 = b$$

Therefore our model will be $A(t) = 450(1.79)^t$.

b. After five years $t = 5$, so we get
$$A(5) = 450(1.79)^5$$
$$A(5) \approx 8269.5$$

Therefore after five years there will be about 8270 worker ants in the colony.

c. After six years $t = 6$, so we get
$$A(6) = 450(1.79)^6$$
$$A(6) \approx 14802.4$$

Our model gives a much higher number of ants than the actual study, which seems to indicate that the colony's growth slowed. This might have occurred because of a limit in food or space available.

From this example and the model found in the concept investigation, you can see that the growth rate becomes part of the base of the exponential. In the concept investigation the growth rate of 4% = 0.04 became part of the 1.04 base of the exponential model. In example 1 we see the 79% = 0.79 also become part of the 1.79 base for the exponential model. In both cases you see that there is a 1 included in the base along with the growth rate. This 1 is there to account for the original population in the calculations while the growth rate accounts for the amount of growth each year. With these examples we can see that the base of the exponential function can be represented by $b = 1 + r$ where r is the growth rate as a decimal. This same base works for exponential decay, except that the rate is considered negative, so the base will end up being smaller than 1.

DEFINITION

Growth rate: The percentage change in a quantity per one unit of time.

$$b = 1 + r$$

$$r = b - 1$$

Where r is a percentage written as a decimal and b is the base of an exponential function of the form $f(x) = a \cdot b^x$.

Using the relationship between a growth rate and the base of an exponential function, we can find a growth rate from a given function. If we are given a starting value for the output and a growth rate, we can also write an exponential model. Since a is the starting value, or vertical intercept, we can easily find a and b to write the model. In Example 1 we knew a starting population of 450 ants and a growth rate of 79% so we could have written the model $A(t) = 450(1.79)^t$ without doing all the work using the two points.

EXAMPLE   2   EXPONENTIAL GROWTH RATE IN A RECOVERING POPULATION

The California sea otter population from 1914 to 1975 was growing at an exponential rate. The population during these years could be modeled by

$$O(t) = 31.95(1.057)^t$$

where $O(t)$ represents the number of California sea otters t years since 1900.

Source: Model derived from data from the National Biological Service.

a.   Use this model to predict the number of California sea otters in 1960.

b.   According to this model what was the growth rate during this time period?

Solution

a.   1960 is represented by $t = 60$, so we get $O(60) = 889$. Thus in 1960 there were approximately 889 California sea otters.

b.   Since the base is 1.057 the growth rate can be found using the formula $r = b - 1$

$$r = 1.057 - 1$$
$$r = 0.057$$

Therefore the California sea otter population was growing at a rate of 5.7% per year.

EXAMPLE ② PRACTICE PROBLEM

According to the CIA World Factbook 2008 the population of Australia can be modeled by

$$P(t) = 19(1.008)^t$$

where $P(t)$ is the population of Australia in millions, t years since 2000.
Source: Model based on information found in the CIA World Factbook 2008.

a.   Use this model to estimate the population of Australia in 2010.

b.   According to this model what is the population of Australia's growth rate?

EXAMPLE 3 FINDING AN EXPONENTIAL MODEL GIVEN A GROWTH RATE

The Gross Domestic Product, GDP, of Germany in 2005 was approximately 2.9 trillion US$ and has been growing by a rate of about 2.5% per year.
Source: Estimated from the CIA World Factbook and www.data.un.org.

a.   Find an equation for a model for the GDP of Germany.

b.   Use your model to estimate the GDP of Germany in 2010.

Solution

a.   First we will define our variables.

t = Time in years since 2005.

G = The Gross Domestic Product of Germany in trillion US$.

Since we know the GDP in 2005 is 2.9 trillion US$, the value of a in our model will be 2.9. The growth rate is 2.5% so we can find the base b using the formula.

$$b = 1 + 0.025$$
$$b = 1.025$$

Therefore our model is $G(t) = 2.9(1.025)^t$.

b.   2010 would be $t = 5$ so we get

$$G(5) = 2.9(1.025)^5$$
$$G(5) \approx 3.281$$

In 2010 the German GDP will be about 3.28 trillion US$.

EXAMPLE ③ PRACTICE PROBLEM

A survey of Boulder, Colorado, residents asked about the optimal size for growth. The results of this survey stated that most residents thought that a growth in population at a rate of 10% per year was desirable. In the year 2000 there were approximately 96,000 people in Boulder.

Source: Census 2000.

a. Find an equation for a model for the population of Boulder, Colorado if the growth rate is 10% per year.

b. Use your model to estimate the population of Boulder, Colorado, in 2002 if the 10% growth rate had been achieved.

EXAMPLE 4 FINDING AN EXPONENTIAL DECAY MODEL GIVEN A GROWTH RATE

In 2008 South Africa had a population of about 43.8 million, but that population was estimated to be decreasing by approximately 0.5% per year.
Source: CIA World Factbook 2008.

a. Find an equation for a model for the population of South Africa.

b. Use your model to estimate the population of South Africa in 2020.

Solution

a. Since the population for 2008 is given, we will use that as our starting year. We will start by defining the variables.

P = The population of South Africa in millions.

t = Time in years since 2008.

Because we have the starting population, we know that $a = 43.8$. The population is decaying, so the growth rate will be represented by a negative number. Therefore, $r = -0.005$, and we have a base of

$$b = 1 - 0.005$$
$$b = 0.995$$

Using these values for a and b, we have the following model.

$$P(t) = 43.8(0.995)^t$$

b. The year 2020 will be represented by $t = 12$, so we get.

$$P(12) \approx 41.24$$

Therefore, the population of South Africa in 2020 will be about 41.24 million.

EXAMPLE ④ PRACTICE PROBLEM

In 2008, Trinidad and Tobago had a population of about 1 million, but that population was estimated to be decreasing by approximately 0.9% per year.
Source: CIA World Factbook 2008.

a. Find an equation for a model for the population of Trinidad and Tobago.

b. Use your model to estimate the population of Trinidad and Tobago in 2020.

COMPOUNDING INTEREST

Exponential equations are found in all kinds of scientific and natural situations as well as a very common business situation: compounding interest. The idea of compounding is that the interest paid to you by a bank for a savings account or certificate of deposit is often compounded daily. Interest that you pay on a loan or credit card balance is also often compounded daily. In the past, many banks would compound interest for savings accounts only annually or monthly to reduce the amount of interest paid to the customers. Understanding the function that is used for compounding interest problems is useful when one is faced with making decisions in many business and financing situations.

For compounding interest problems we use the formula

$$A = P\left(1 + \frac{r}{n}\right)^{nt}$$

where the variables represent the following.

A = The amount in the account

P = The principal (the amount initially deposited)

r = The annual interest rate written as a decimal

n = The number of times the interest is compounded in one year

t = The time the investment is made for, in years

To use this formula you need to identify each quantity and substitute it into the appropriate variable. Then be sure to use the order of operations when calculating the result.

CONCEPT INVESTIGATION 2 WHAT IS THE DIFFERENCE BETWEEN SIMPLE AND COMPOUND INTEREST?

In this investigation, we will compare two investments, one that earns simple interest and one that earns annually compounded interest. With simple interest, we use the formula $A = P + Prt$ to calculate the account balance A after P dollars have been invested for t years. With annually compounding interest, we will use the formula above, and n will be 1, since the interest will compound only once a year. Since $n = 1$, the formula simplifies to $A = P(1 + r)^t$.

If we assume a \$100 investment earning 10% interest, complete the following table of calculations.

Years	Simple Interest	Annually Compounding Interest
0	100	100
1	$100 + 100(0.10)(1) = 110$	$100(1 + 0.10)^1 = 110$
2		
5		
10		
20		
50		

How do simple interest and compounding interest compare?

What would happen if you increased the number of compounding periods per year?

◆

EXAMPLE 5 USING THE COMPOUNDING INTEREST FORMULA

If $5000 is invested in a savings account that pays 3% annual interest compounded daily, what will the account balance be after four years?

Solution

First identify each given quantity and decide what you are solving for.

$5000 is the initial deposit, so that is represented by $P = 5000$.

3% is the annual interest rate, so it is represented by $r = 0.03$.

"Compounded daily" tells us that there will be 365 compounds in one year, so we have $n = 365$.

Four years is the time the investment is made for, so it is represented by $t = 4$.

We are asked to find the amount in the account, so we need to solve for A. All of this gives us the following equation.

$$A = 5000\left(1 + \frac{0.03}{365}\right)^{365(4)}$$

$$A = 5000(1 + 0.000082192)^{1460}$$

$$A = 5000(1.1275)$$

$$A \approx 5637.46$$

EXAMPLE PRACTICE PROBLEM

An initial deposit of $30,000 is placed in an account that earns 8% interest. Find the amount in the account after 10 years if the interest is compounded:

a. monthly   **b.** weekly   **c.** daily

One other type of compounding growth is discussed in some areas of business and finance. Continuous compounding is the idea that the interest is always being compounded on itself, resulting in a higher return. In dealing with this type of problem, the compounding interest equation that we have been using will no longer work, since n would have to be infinity. Use the following concept investigation to discover what happens to this formula when n goes to infinity.

CONCEPT INVESTIGATION 3 WHAT ABOUT CONTINUOUS COMPOUNDING?

We will assume we are investing $1 at 100% interest for 1 year and let the number of compounds per year increase toward infinity. Complete the following table using your calculator. If you use the table feature you will need to move your cursor to the Y1 column to see more decimal places.

n	$A = 1\left(1 + \dfrac{1}{n}\right)^n$
1	2
100	2.7048138
1,000	
10,000	
100,000	
1,000,000	
5,000,000	

Above the division button ➗ is the number e. Press [2nd] [➗] [ENTER] and write down the number you get.

How does the number e compare with the number you found as n went toward infinity?

e is a special number that occurs often in nature and science as well in compounding interest situations.

DEFINITION

The number e: An irrational number represented by the letter e.

$$e \approx 2.718281828$$

If you look carefully on your calculator, you will find a button with e^x on the left side just above the [LN] button. If you look at the graph of the following exponential functions, you can see that e lies between 2 and 3.

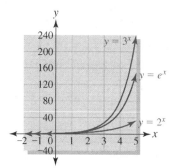

For now we will use e for continuously compounding interest problems and we will see it again in the next chapter.

When compounding interest continuously the compounding interest equation ends up being transformed into the equation

$$A = Pe^{rt}$$

where A is the account balance after the principal P has invested at an annual rate r compounded continuously for t years. In this equation, $e \approx 2.718281828$, and you can use the e^x key on your calculator to perform the calculations.

EXAMPLE 6 CONTINUOUSLY COMPOUNDING INTEREST

If $2000 is invested in a savings account that pays 5% annual interest compounded continuously, what will the account balance be after seven years?

Solution

We know the following values.

$$P = 2000 \qquad r = 0.05 \qquad t = 7$$

A is the missing quantity that we are being asked to solve for.

$$A = 2000e^{0.05(7)}$$

$$A \approx 2838.14$$

Therefore, after seven years, this account will have $2,838.14 in it.

EXAMPLE 6 PRACTICE PROBLEM

An initial deposit of $80,000 is placed in an account paying 5% annual interest compounded continuously. What will the account balance be after 12 years?

5.5 Exercises

For Exercises 1 through 10 find the growth rate of the given exponential model.

1. $f(x) = 45(1.03)^x$

2. $f(x) = 20(1.07)^x$

3. $g(x) = 130(1.25)^x$

4. $h(x) = 3(1.46)^x$

5. $f(x) = 2(3.5)^x$

6. $g(x) = 7(4.75)^x$

7. $h(x) = 25(0.95)^x$

8. $f(x) = 150(0.92)^x$

9. $f(x) = 750(0.36)^x$

10. $g(x) = 5000(0.001)^x$

11. The white-tailed deer population in the northeastern United States has been growing since the early 1980s. The population of white-tailed deer can be modeled by

$$W(t) = 1.19(1.08)^t$$

where $W(t)$ represents the number of white-tailed deer in millions in the Northeast t years since 1980.

Source: Model derived from data from the National Biological Service.

   a. Estimate the white-tailed deer population in 1990.

   b. According to this model what is the growth rate of the white-tailed deer population?

   c. What could cause this population to stop growing at this rate?

12. According to United Nations records, the world population in 1975 was approximately 4 billion. Statistics indicate that the world population since World War II has been growing exponentially. If we assume exponential growth, the world population can be modeled by

$$P(t) = 4(1.019)^t$$

where $P(t)$ is the world population in billions and t is the time in years since 1975.

   a. Estimate the world population in 2015.

   b. According to this model, what is the growth rate of the world population?

13. The population of the Virgin Islands can be modeled by

$$V(t) = 108(0.73)^t$$

where $V(t)$ is the population of the Virgin Islands in thousands t years since 2008.

   a. Estimate the population of the Virgin Islands in 2015.

   b. According to this model, what is the decay rate of the Virgin Islands' population?

14. Dr. Arnd Leike, a professor of physics at the University of Munich in Germany, won a 2002 Ig Nobel Prize in Physics for his investigation into the "exponential decay" of a beer's head (the foam on top when poured). After testing several beers for the rate of decay of the head, he came to the conclusion that a beer could be identified by its unique head decay rate. The following functions are for three of the beers that Dr. Leike tested.

$$E(s) = 16.89(0.996)^s$$
$$B(s) = 13.25(0.991)^s$$
$$A(s) = 13.36(0.993)^s$$

where $E(s)$ represents the height of the head of an Erdinger beer, $B(s)$ is the height of the head of a Budweiser Budvar, and $A(s)$ is the head height of an Augustinerbrau. All head heights are in centimeters s seconds after being poured.

Source: European Journal of Physics, Volume 23, 2002.

   a. Use these models to estimate the height of each head at the end of the pour.

   b. Use these models to estimate the height of each head after 200 seconds.

   c. Use these models to determine the decay rate of each head of beer.

15. A 13-year research study found that the humpback whale population off the coast of Australia was increasing at a rate of about 14% per year. In 1981 the study estimated the population to be 350 whales.

a. Find an equation for a model for the humpback whale population.

b. Give a reasonable domain and range for the model.

c. Estimate the humpback whale population in 1990.

16. The population of Egypt has been growing at a rate of 1.68% per year. In 2008 the population of Egypt was about 81.7 million.

Source: CIA World Factbook 2008.

a. Assuming exponential growth, find an equation for a model for the population of Egypt.

b. Estimate Egypt's population in 2020.

17. The population of Denmark has been growing at a natural rate of 0.295% per year. In 2008 the population of Denmark was about 5.5 million.

Source: CIA World Factbook 2008.

a. Assuming exponential growth, find an equation for a model for the population of Denmark.

b. Estimate Denmark's population in 2020.

18. The sandhill crane is a migratory bird in the United States that is being watched for its population trends. The National Biological Service estimates the population's present trend to be a growth of approximately 4.3% per year. In 1996 the estimated population was 500,000.

Source: National Biological Service.

a. Assuming that this trend continues, find an equation for a model for the sandhill crane population.

b. Estimate the population in 2002.

Use the bank brochure to answer Exercises 19–34.

New Savings Options!

Standard CD:
Minimum balance required $5000.00, 3% interest

Silver CD:
Minimum balance required $10,000, 5% interest

Gold CD:
Minimum balance required $100,000, 7% interest

Platinum CD:
Minimum balance required $500,000, 9% interest

19. An initial deposit of $10,000 is placed in a Silver CD. Find the amount in the account after 10 years if the interest is compounded monthly.

20. An initial deposit of $10,000 is placed in a Silver CD. Find the amount in the account after 10 years if the interest is compounded weekly.

21. An initial deposit of $10,000 is placed in a Silver CD. Find the amount in the account after 10 years if the interest is compounded daily.

22. An initial deposit of $10,000 is placed in a Silver CD. Find the amount in the account after 10 years if the interest is compounded continuously.

23. An initial deposit of $500,000 is placed in a Platinum CD. Find the amount in the account after 10 years if the interest is compounded monthly.

24. An initial deposit of $500,000 is placed in a Platinum CD. Find the amount in the account after 10 years if the interest is compounded daily.

25. An initial deposit of $500,000 is placed in a Platinum CD. Find the amount in the account after 10 years if the interest is compounded hourly.

26. An initial deposit of $500,000 is placed in a Platinum CD. Find the amount in the account after 10 years if the interest is compounded continuously.

27. An initial deposit of $100,000 is placed in a Gold CD. Find the amount in the account after 20 years if the interest is compounded quarterly.

28. An initial deposit of $100,000 is placed in a Gold CD. Find the amount in the account after 20 years if the interest is compounded monthly.

29. An initial deposit of $100,000 is placed in a Gold CD. Find the amount in the account after 20 years if the interest is compounded daily.

30. An initial deposit of $100,000 is placed in a Gold CD. Find the amount in the account after 20 years if the interest is compounded continuously.

31. An initial deposit of $5000 is placed in a Standard CD. Find the amount in the account after five years if the interest is compounded quarterly.

32. An initial deposit of $5000 is placed in a Standard CD. Find the amount in the account after five years if the interest is compounded monthly.

33. An initial deposit of $5000 is placed in a Standard CD. Find the amount in the account after five years if the interest is compounded weekly.

34. An initial deposit of $5000 is placed in a Standard CD. Find the amount in the account after five years if the interest is compounded continuously.

Chapter 5 Summary

Section 5.1 Introduction to Exponentials

- An **exponential function** can be written in the form $f(x) = a \cdot b^x$, where a and b are real numbers such that $a \neq 0$, $b > 0$, and $b \neq 1$.

- The constant b is called the **base** of the exponential function.

- The constant $f(0) = a$, and thus $(0, a)$ is the **vertical intercept** for the graph.

- **Exponential growth and decay** are two examples of exponential patterns that we find in the world.

- Exponentials occur when a quantity is being repeatedly multiplied by the same constant.

- When looking for an exponential pattern, try to find the quantity used repeatedly as the multiplier. Once you do so, it will help you to find the base b.

EXAMPLE

A biologist is studying the growth of 50 salmonella bacteria. When given the right food source and temperature range, these bacteria double in number every 15 minutes.

a. Find an equation for a model for the number of salmonella bacteria after h hours.

b. Use your model to estimate the number of salmonella bacteria after 3 hours.

c. Estimate graphically when the number of salmonella bacteria will reach 500,000.

Solution

a.

> h = The number of hours since starting with 50 salmonella bacteria.
>
> S = The number of salmonella bacteria.

Because the bacteria double every 15 minutes, they will double 4 times every hour. By building a table, we can find a pattern.

h	0	1	2	3	h
S	50	$50(2)^4$	$50(2)^8$	$50(2)^{12}$	$50(2)^{4h}$

From the pattern we have the model $S(h) = 50(2)^{4h}$.

b. $S(3) = 50(2)^{4(3)} = 204800$, so after 3 hours there will be about 204,800 salmonella bacteria.

c. Graphing the function and 500,000, we can trace to find an estimate.

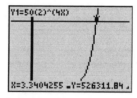

Therefore there will be about 500,000 salmonella bacteria after about 3.3 hours.

Section 5.2 Solving Equations Using Exponent Rules

- When **solving simple exponential equations,** try to get the bases on each side of the equation the same and set the exponents equal to one another.

- When **solving power equations,** get the variable being raised to a power by itself and use the reciprocal expo-

nent of both sides to undo it. Remember to use the plus/minus symbol whenever the reciprocal has an even denominator.

EXAMPLE 2

Solve the following equations.

a. $2^x = 16$

b. $3x^7 = 126$

Solution

a. $2^x = 16$

$$2^x = 2^4$$
$$x = 4$$

b. $3x^7 = 126$
$$x^7 = 42$$
$$(x^7)^{\frac{1}{7}} = (42)^{\frac{1}{7}}$$
$$x = 1.706$$

Section 5.3 Graphing Exponential Functions

- The graph of an exponential function is affected by the values of a and b: $f(x) = a \cdot b^x$.

- The vertical intercept for the graph of an exponential function is $(0, a)$.

- If a is negative, the graph of the function will be reflected over the horizontal axis.

- If $a > 0$ and $0 < b < 1$, then the graph will be decreasing. The smaller the value of b, the steeper the decrease.

- If $a > 0$ and $b > 1$, than the graph will be increasing. The larger the value of b, the steeper the increase.

- The domain of an exponential function without a context will be all real numbers.

- The range of an exponential function without a context will be either the positive real numbers or the negative real numbers depending on the sign of a.

x	f(x)
3	6.9
6	11.9
9	20.6

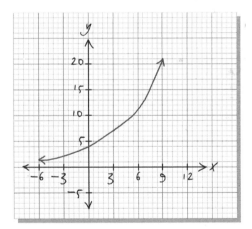

EXAMPLE 3

Sketch a graph by hand of the function $f(x) = 4(1.2)^x$.

Solution

a. $a = 4$, so the vertical intercept is $(0, 4)$, and the graph will be above the horizontal axis. The base $b = 1.2$, so this graph is increasing, showing exponential growth. By trying a few input values, we get a table of points; using these points to plot the graph, we get the following graph

EXAMPLE 4

Give the domain and range of the function $f(x) = -2(1.7)^x$.

Solution

Domain: all real numbers. Range: $(-\infty, 0)$.

Section 5.4 Finding Exponential Models

* To find an exponential model, follow these steps.

  1. Define the variables and adjust the data.

  2. Create a scatterplot.

  3. Select a model type.

  4. Exponential Model: Pick two points, write two equations using the form $f(x) = a \cdot b^x$ and solve for b.

  5. Use b and one equation to find a.

  6. Write the equation of the model using function notation.

  7. Check your model by graphing it with the scatterplot.

* If the data you are modeling show the vertical intercept, you can use that intercept for a and one equation to help find b.

* In the context of an application the domain of an exponential model expands beyond the data. Be careful to avoid inputs that will cause model breakdown to occur. The range will be the lowest to highest points on the graph within the domain.

EXAMPLE 5

The gross profit for eBay, Inc. is given in the table.

Year	Gross Profit (in millions of $)
1997	33
1998	70
1999	167
2000	336
2002	1000.2
2003	1749

Source: SEC Financial Filings from eBay, Inc., 2004.

a. Find an equation for a model for the data.

b. Estimate the gross profit for eBay, Inc. in 2001.

c. Give a reasonable domain and range for your model.

Solution

a. P = The gross profit for eBay, Inc. in millions of dollars
t = Time in years since 1990

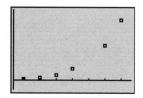

We can use the points (8, 70) and (12, 1000.2).

$$70 = a \cdot b^8 \qquad 1000.2 = a \cdot b^{12}$$

$$\frac{1000.2}{70} = \frac{a \cdot b^{12}}{a \cdot b^8}$$

$$14.289 = b^4$$

$$1.94 = b$$

$$70 = a \cdot (1.94)^8$$

$$0.349 = a$$

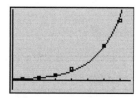

So we have the model $P(t) = 0.349(1.94)^t$.

b. $P(11) = 511.26$. eBay, Inc. had a gross profit of about $511.3 million in 2001.

c. Domain: [6, 13], range: [18.6, 1924.2]. Beyond this domain, model breakdown may occur.

Section 5.5 Exponential Growth and Decay Rates and Compounding Interest

- The **growth rate, r,** is the percentage change in a quantity per one unit of time.

$$b = 1 + r$$

$$r = b - 1$$

  where b is the base of the exponential function of the form $f(x) = a \cdot b^x$.

- The **compound interest** formula is

$$A = P\left(1 + \frac{r}{n}\right)^{nt}$$, where A is the account balance,

  P is the principal invested, r is the interest rate (as a decimal), n is the number of times the account compounds per year, and t is the time in years of the investment. This formula works if n is any finite number.

- If the interest rate is **compounding continuously,** we use the formula $A = Pe^{rt}$, where A is the account balance, P is the principal invested, r is the interest rate (as a decimal), and t is the time in years of the investment.

EXAMPLE 6

If a population of 200 is growing at 3% per year, find an exponential model for the size of the population after t years.

Solution

Let P be the size of the population after t years.
Since the population starts with 200, we know that $a = 200$. We are given a growth rate, so the base is

$$b = 1 + 0.03$$

$$b = 1.03$$

Therefore the model would be

$$P(t) = 200(1.03)^t$$

EXAMPLE 7

$5000 is invested in an account for 30 years.

a. Find the balance in the account if the money earns 4% interest compounded monthly.

b. Find the balance in the account if the money earns 4% interest compounded continuously.

Solution

a. $A = 5000\left(1 + \dfrac{0.04}{12}\right)^{12(30)} = 16567.49$.

   The account will have \$16,567.49 after 30 years if compounded monthly.

b. $A = 5000 \cdot e^{0.04(30)} = 16600.58$. The account will have \$16,600.58 after 30 years if compounded continuously.

Chapter 5 Review Exercises

1. The population of Africa is estimated to be doubling about every 28 years. The population of Africa in 1996 was about 731.5 million.

   Source: U.S. Department of Commerce, Bureau of the Census.

   a. Using this information, find an equation for a model for the population of Africa.

   b. Using your model, estimate the population of Africa in 2005. **[5.1]**

2. The flu is traveling through a large city at an alarming rate. Every day the number of people who have flu symptoms is 3 times what it was the previous day. Assume that one person started this flu epidemic.

   a. Find an equation for a model for the number of people with flu symptoms after d days.

   b. Use your model to estimate the number of people with flu symptoms after two weeks. **[5.1]**

3. Polonium-210 has a half-life of 138 days.

   a. Find an equation for a model for the percent of a polonium-210 sample left after d days.

   b. Estimate the percent of polonium-210 left after 300 days. **[5.1]**

4. Thorium-228 has a half-life of 1.9 years.

   a. Find an equation for a model for the percent of a thorium-228 sample left after t years.

   b. Estimate the percent of thorium-228 left after 50 years. **[5.1]**

5. Find an exponential model for the data. **[5.1]**

x	$f(x)$
0	5000
1	4000
2	3200
3	2560
4	2048

6. Find an exponential model for the data. **[5.1]**

x	$f(x)$
0	0.2
1	1.52
2	11.53
3	87.58
4	665.05

   For Exercises 7 through 12 solve the given equation.

7. $4^x = 1024$ **[5.2]**

8. $5(3^x) = 10935$ **[5.2]**

9. $2x^5 + 7 = 1057.4375$ **[5.2]**

10. $-4x^8 + 2768 = -6715696$ **[5.2]**

11. $2^x = \dfrac{1}{32}$ **[5.2]**

12. $(-3)^x = \dfrac{1}{81}$ **[5.2]**

   For Exercises 13 through 16, sketch the graph of the function and state any information you know from the values of a and b.

13. $f(x) = 5(1.3)^x$ **[5.3]**

14. $g(t) = -0.25(0.5)^t$ **[5.3]**

15. $f(x) = 200(0.2)^x$ **[5.3]**

16. $g(t) = -150(0.6)^t$ **[5.3]**

17. Give the domain and range of the function in exercise 13. **[5.3]**

18. Give the domain and range of the function in exercise 14. **[5.3]**

19. The number of nonstrategic nuclear warheads stockpiled by the United States during the 1950s is given in the chart. **[5.4]**

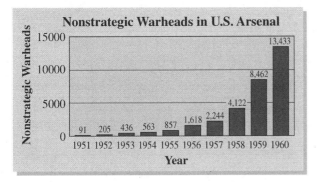

Nonstrategic Warheads in U.S. Arsenal

Source: Natural Resources Defense Council.

a. Find an equation for a model for these data.

b. Give a reasonable domain and range for your model.

c. Estimate the number of nonstrategic warheads stockpiled by the United States in 1965.

20. The following data were collected during an experiment designed to measure the decay of the radioactive material barium-137 . Every 60 seconds the number of counts per minute was taken from a Geiger counter. The readings are recorded below.

Time (in seconds)	Counts per Minute
0	3098
60	2270
120	1856
180	1436
240	1220
300	914
360	596

a. Find an exponential model for these data.

b. Estimate the number of counts per minute this sample would have after 10 minutes.

c. Give a reasonable domain and range for this model. **[5.4]**

21. The number of Franklin's gulls in North America has been in significant decline in the past several years. The population of Franklin's gulls can be modeled by

$$F(t) = 750(0.9405^t)$$

where $F(t)$ represents the number of Franklin's gulls in thousands in North America t years since 2000.

Source: Model derived from data from the National Biological Service.

a. Estimate the Franklin's gull population in 2005.

b. According to the model what is the rate of decay for the Franklin's gull population?

c. What might cause this population to stop declining at this rate? **[5.5]**

22. The population of the European Union was about 491 million in 2008 and was growing at a rate of approximately 0.12% per year.

Source: CIA World Factbook 2008.

a. Find an equation for a model of the population of the European Union.

b. Estimate the population of the European Union in 2020. **[5.5]**

23. What will an investment of $7000 be worth in 10 years if it is deposited in an account that pays 6% interest compounded monthly? **[5.5]**

24. If an investment of $100,000 is deposited in an account that pays 8.5% interest compounded continuously, what will the account balance be after 25 years? **[5.5]**

25. If $50,000 is deposited in an account paying 4% interest compounded daily, what will the account balance be after 15 years? **[5.5]**

Chapter 5 Test

1. At 12 noon there are 5 million bacteria on the bathroom door handle. Under these conditions the number of bacteria on the handle is doubling every hour.

   a. Find an equation for a model for the number of bacteria on the bathroom door handle h hours after 12 noon.

   b. Estimate the number of bacteria on the handle at 6 p.m.

   c. Find a new model for this situation if the bacteria are doubling every 15 minutes.

2. Thallium-210 has a half-life of 1.32 minutes.

   a. Find an equation for a model for the percent of a thallium-210 sample left after m minutes.

   b. Estimate the percent of thallium-210 left after 15 minutes.

3. How much will an investment of $1000 be worth after 12 years if it is deposited in an account that pays 3.5% interest compounded monthly?

4. How much will an investment of $40,000 be worth after 35 years if it is deposited in an account that pays 6% compounded continuously?

5. Use the graph of the exponential to answer the following.

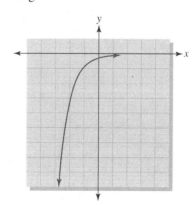

   a. Is the value of a positive or negative?

   b. Is the value of b less than or greater than 1?

6. The number of CD singles shipped in the mid-1990s is given in the following table.

Year	CD Singles (in millions)
1993	7.8
1994	9.3
1995	21.5
1996	43.2
1997	66.7

Source: Statistical Abstract 2001.

   a. Find an equation for a model for these data.

   b. Give a reasonable domain and range for this model.

   c. Estimate the number of CD singles shipped in 2000.

For Exercises 7 and 8, sketch the graph of the function and state any information you know from the values of a and b. Give the domain and range of each function.

7. $f(x) = -2(3.1)^x$

8. $g(t) = 40(0.8)^t$

For Exercises 9 and 10, solve the given equation.

9. $2(3^x) - 557 = -503$

10. $4.5x^6 - 2865 = 415.5$

Chapter 5 Projects

HOW HOT IS THAT WATER?

In this project you will test the temperature of a cup of hot water over time. Using a Texas Instruments CBL unit and the temperature probe, measure the room temperature at the beginning of the experiment. Then put the probe into the hot water, allowing the probe to warm up for a couple of seconds before you start to collect temperature data from the water. Try to collect data for at least 45 minutes or longer. Be sure to keep the cup in a location that will not get too much of temperature change during the experiment (away from air conditioning vents or open doors).

Write up

a. What was the room temperature at the beginning of the experiment?

b. Create a scatterplot of the data on the calculator or computer and print it out or neatly by hand on graph paper.

c. As more and more time goes by, what temperature is the water approaching? Why?

d. What kind of model would best fit these data? Explain your reasons.

e. Complete the following steps to model the data.

1. Adjust the data by subtracting the room temperature from each temperature value.

2. Find an equation for a model for the data.

3. Add the room temperature to your model.

4. Graph the model with the original data you collected.

f. What is the vertical intercept of your model and what does it represent in this experiment?

g. What are a reasonable domain and range for your model? (Remember to consider any restrictions on the experiment.)

NOW THAT'S A LOT OF RABBITS!

Research Project

One or more people

What you will need:

• Find information regarding the rabbit problem in Australia.

In this project you will investigate the rabbit problem in Australia. In 1859, 24 wild rabbits were brought to Australia and released. This original population of 24 rabbits grew to an estimated 600,000,000 rabbits by 1950. This was, and still is, devastating to the Australian economy and natural resources.

Write up

a. Research the rabbit problem in Australia and describe the reasons for such a large growth in the number of rabbits.

b. Assuming that the growth in the rabbit population is exponential over these years, Find an equation for a model to describe the number of rabbits in Australia during this time period.

c. Did the population of rabbits in Australia continue to grow at such a fast pace after 1950? Why or why not?

d. What are a reasonable domain and range for your model?

e. What tactics have been used to try to stop the rabbit population from growing? How successful were these tactics?

FIND YOUR OWN EXPONENTIAL

Research Project

One or more people

What you will need:

• Find data for a real-world situation that can be modeled with an exponential function.

• You might want to use the Internet or library. Statistical abstracts and some journals and scientific articles are good resources for real world data.

In this project you are given the task of finding data for a real-world situation for which you can find an exponential model. You may use the problems in this chapter to get ideas of things to investigate, but your data should not be discussed in this textbook.

Write up

a. Describe the data you found and where you found it. Cite any sources you used.

b. Create a scatterplot of the data on the calculator or computer and print it out or draw it neatly by hand on graph paper.

c. Describe what about the data you collected and/or the situation led you to use an exponential model.

d. Find an equation for a model to fit the data.

e. What is the vertical intercept of your model and what does it represent in the context?

f. What are a reasonable domain and range for your model?

g. Use your model to estimate an output value of your model for an input value that you did not collect in your original data.

h. Use your model to estimate the input value for which your model will give you a specific output value that you did not collect in your original data.

Logarithmic Functions

- Describe and find inverses.
- Use the horizontal line test to verify one-to-one functions.
- Understand logarithms and their inverse relationship to exponentials.
- Find inverses of exponential functions.
- Graph logarithm functions.
- Find the domain and range of logarithm functions.
- Use the properties of logarithms to simplify and expand logarithm expressions.
- Use the change of base formula.
- Solve exponential equations.
- Solve applications of compounding interest formulas.
- Solve Logarithm equations.
- Use logarithms in applications.

  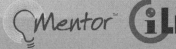

Introduction to Inverses

INTRODUCTION TO INVERSE FUNCTIONS

In chapter 5 we investigated exponential functions but have not yet solved for the input variable because it is in the exponent. In this section we will discuss the concept of an **inverse function** and build the basic rules so that we can define the inverse for exponential functions in the next section. These will be used throughout the rest of this chapter to solve exponential functions.

 CONCEPT INVESTIGATION | WHICH WAY DO YOU WANT TO GO?

Today's Weather

London

 Cloudy with moderate showers at times
High 21°C
Low 15°C

Birmingham

 Cloudy with light showers
High 20°C
Low 13°C

Temperature is measured in different ways around the world. In the United States we commonly measure temperature in degrees Fahrenheit, but in most other countries temperature is measured in degrees Celsius. This means that people often have to switch a given temperature from one unit to another. This is especially true for people who travel to different countries and get temperatures in different units. The function

$$C(F) = \frac{5}{9}(F - 32)$$

has the temperature in degrees Fahrenheit as the input value and the temperature in degrees Celsius as the output value.

a. If you know that the temperature is 68°F outside, calculate the temperature in degrees Celsius.

   If you were given the temperature in degrees Celsius, it would be convenient to have a function that had an input variable that took degrees Celsius and gave you out the value in degrees Fahrenheit. To find such a function, we can simply solve the above function for *F*.

b. Solve the function $C = \frac{5}{9}(F - 32)$ for *F*.

c. Use the function that you just found to change 20 degrees Celsius into degrees Fahrenheit.

   The function that you found in part b undoes the operations that the original given function had done. These functions are an example of what we call *inverses*. When one function "undoes" the operations of the other function, you are looking at an inverse. Finding inverses for linear equations is not a difficult process.

> **FINDING AN INVERSE FUNCTION IN A REAL-WORLD PROBLEM**
>
> **1.** Write without function notation.
> **2.** Solve for the "other" variable, i.e. the original input variable. This will make it the new output variable.
> **3.** Rewrite in function notation.

EXAMPLE ▮ FINDING AND USING AN INVERSE IN AN APPLICATION

In Chapter 1 we looked at a function to calculate the cost to rent a 10-foot truck from U-Haul.

$$U(m) = 0.79m + 19.95$$

where U represented the cost in dollars to rent a 10-foot truck from U-Haul and drive it m miles.

a. What are the input and output variables for this function? How would this function be used?

b. Find the inverse of this function.

c. What are the input and output variables for the inverse? How would this function be used?

d. Find the cost to travel 100 miles in this truck.

e. Find the number of miles you can travel in this truck for a cost of $150.

Solution

a. In this function, m miles is the input variable, and U dollars is the output variable. If you wanted to know the cost to drive m miles, you would substitute in the number of miles and get out the cost in dollars.

b. To find the inverse of this function, we basically want to make m the output and U the input. To do this, it is easiest not to use function notation until the end of the process.

$$U(m) = 0.79m + 19.95$$
$$U = 0.79m + 19.95 \qquad \text{Write without function notation.}$$
$$U - 19.95 = 0.79m \qquad \text{Solve for } m.$$
$$\frac{U - 19.95}{0.79} = m \qquad \text{Divide by 0.79.}$$
$$1.27U - 25.25 = m$$
$$m(U) = 1.27U - 25.25 \qquad \text{Rewrite in function notation.}$$

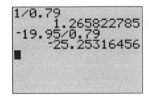

c. For the inverse the input variable is now U dollars, and the output is now m miles. This function would be used if you had a cost in mind and wanted to know how many miles you could drive the truck.

d. The number of miles is given, so it would be best to use the original function where the input is m = number of miles.

$$U(m) = 0.79m + 19.95$$

$$U(100) = 0.79(100) + 19.95$$

$$U(100) = 98.95$$

Therefore it cost $98.95 to rent this truck and drive it 100 miles.

e. The cost is the given quantity, so it is easiest to use the inverse function where the input is U = the cost of the rental.

$$m(U) = 1.27U - 25.25$$

$$m(150) = 1.27(150) - 25.25$$

$$m(150) = 165.25$$

Therefore you can drive about 165 miles for a rental cost of $150.

Inverses basically take the inputs and outputs of one function and reverse their roles. The input of one function will become the output of the inverse and vice versa. This also means that the domain and range of an inverse are simply the domain and range of the original function—but reversed. The domain of the function is the range of its inverse, and the range of the function is the domain of the inverse.

In the U-haul example the domain for the function may be something like $0 \leq m \leq 400$ miles driven for the day. With this domain the range would be the lowest to highest cost for those miles giving us $19.95 \leq U \leq 335.95$. Since the inverse of this function has the cost as the input variable, the domain will be the costs $19.95 \leq U \leq 335.95$ and the range will be the possible miles driven $0 \leq m \leq 400$. Here you can see that the domain and range for the inverse simply switched roles.

DOMAIN AND RANGE OF INVERSE FUNCTIONS

The domain and range of inverse functions will switch roles.

Inverse Functions

$U(m)$ $m(U)$

Domain: $0 \leq m \leq 400$ Domain: $19.95 \leq U \leq 335.95$

Range: $19.95 \leq U \leq 335.95$ Range: $0 \leq m \leq 400$

It is important in a real-world problem that you keep the units and definitions of each variable intact so that their meanings do not get lost. The definition of the variables is what will control the meaning of the input and outputs of the function and its inverse. The U, in the U-haul function and its inverse, represents the cost in dollars and the m, in both functions, represents miles driven.

EXAMPLE PRACTICE PROBLEM

A team of engineers is trying to pump down the pressure in a vacuum chamber. They know that the following equation represents the pressure in the chamber.

$$P(s) = 35 - 0.07s$$

where P is the pressure in pounds per square inch (psi) of a vacuum chamber after s seconds.

a. Find the inverse for this function.

b. Use the inverse function to estimate the time it will take to pump down this vacuum chamber to 5 psi.

c. If the original function had a domain of $0 \le s \le 500$ and range of $0 \le P \le 35$, what are the domain and range of the inverse?

In working with inverses in a real-world situation, the meanings of the variables play an important role in making the equation apply to the context of the problem. In problems without a context we do not define the variables, so most often x is the input variable for a function and $f(x) = y$ is the output for the function. When they are finding an inverse function, because mathematicians want the input variable of all functions to be x, they will interchange the variables. To keep the original function and its inverse distinct, mathematicians use different notation when working with inverses without a context. If the original function is $f(x)$, the inverse will have a negative 1 placed above the f to indicate that it is the inverse of the function $f(x)$. Therefore the inverse function will be written as $f^{-1}(x)$, and read "f inverse of x."

FINDING AN INVERSE FUNCTION IN A PROBLEM WITHOUT A CONTEXT

1. Write without function notation by replacing $f(x)$ with y.

2. Solve for x, the original input variable. This will make it the new output variable.

3. Interchange the variables x and y.

4. Rewrite in function notation. Use $f^{-1}(x)$ to designate it as the inverse. Note that the -1 in the inverse notation, $f^{-1}(x)$, is not an exponent. It simply indicates an inverse in function notation.

When you are finding an inverse, steps 2 and 3 above can be swapped and it will not affect the inverse function.

EXAMPLE FINDING AN INVERSE FUNCTION

Find the inverse for the following functions

a. $f(x) = 2x - 8$ **b.** $g(x) = \frac{1}{3}x + 7$

Solution

a. To find the inverse, we still want to reverse the roles of the two variables involved. This is most easily done without using function notation and then going back to it at the end of the problem.

Step 1 Write without function notation by replacing $f(x)$ with y.

$$f(x) = 2x - 8$$
$$y = 2x - 8$$

Step 2 Solve for x. (This will reverse the role of the two variables.)

$$y = 2x - 8$$
$$y + 8 = 2x$$
$$\frac{y + 8}{2} = \frac{2x}{2} \qquad \text{Divide both sides by 2.}$$
$$0.5y + 4 = x$$

Step 3 Interchange the variables x and y. We do this only when there is no context because the input is usually denoted by x.

$$0.5x + 4 = y$$

Step 4 Rewrite in function notation. Use $f^{-1}(x)$ to designate it as the inverse.

$$f^{-1}(x) = 0.5x + 4$$

b. We will follow the same 4 steps.

$$g(x) = \frac{1}{3}x + 7$$

$$y = \frac{1}{3}x + 7 \qquad \text{Write without function notation.}$$

$$y - 7 = \frac{1}{3}x \qquad \text{Solve for } x.$$

$$3(y - 7) = 3\left(\frac{1}{3}x\right)$$

$$3y - 21 = x$$

$$3x - 21 = y \qquad \text{Interchange the variables } x \text{ and } y.$$

$$g^{-1}(x) = 3x - 21 \qquad \text{Put back into function notation using the -1.}$$

EXAMPLE ② PRACTICE PROBLEM

Find the inverse for the following functions.

a. $f(x) = -4x + 9$ **b.** $g(t) = 2.5t - 3.5$

When you are working with linear functions, the "undoing" process of inverses can be seen clearly. If we look at a function and its inverse we can see that the inverse uses the opposite operations but they are done in reverse order.

$$f(x) = 2x + 1 \qquad\qquad f^{-1}(x) = \frac{1}{2}(x - 1)$$

Input x

x

Multiply by 2

$2x$

Add 1

$2x + 1$

$$f(x) = 2x + 1$$

Input x

x

Subtract 1 (to undo the addition)

$x - 1$

Divide by 2 (to undo the multiplication)

$$\frac{(x - 1)}{2} = \frac{1}{2}(x - 1)$$

$$f^{-1}(x) = \frac{1}{2}(x - 1)$$

Notice that the steps are in reverse order and undo each other.

ONE-TO-ONE FUNCTIONS

Recall that a function was defined as a relation where each input x results in exactly one output y. In an inverse of a function the variables x and y are swapped so this relationship of each input to exactly one output must work in reverse for the inverse to be a function. Therefore inverse functions are possible only if each output of an original function can be brought back to a single input value that it came from. When an output has more than one input it is associated with it becomes a problem. An example would be the function $f(x) = x^2$, which has the same output of 4 when $x = 2$ as when $x = -2$, so if you want to go back from an output of 4 to the original input, $x = 2$ and $x = -2$ are both options, and choosing which one it should be is a problem. Because we want an inverse to be a function, this problem of one output coming from two different inputs, eliminates the possibility of $f(x) = x^2$ having an inverse function. If an inverse is not a function we are not interested. Many types of functions do not have inverse functions.

Functions that have only one input for any one output are called **one-to-one functions**; that is, each input goes to exactly one output, and each output comes from exactly one input. A function that is one-to-one will have an inverse function. A one-to-one function is easiest to identify by using its graph. If a function is one-to-one, it will pass **the horizontal line test** much as a function must pass the vertical line test discussed earlier.

DEFINITIONS

One-to-one function: A function in which each input corresponds to only one output **and** each output corresponds to only one input.

The horizontal line test for a one-to-one function: If any horizontal line intersects the graph of a function at most once, then that graph is a one-to-one function.

EXAMPLE 3 USING THE HORIZONTAL LINE TEST

Determine whether each of the following functions is a one-to-one function.

a. $f(x) = -2.4x + 9$ **b.** $f(x) = 2.5x^2 + 3x - 5$

c. $g(t) = 3.4(1.4^t)$ **d.** $h(t) = 5$

Solution

Each of the following functions can be graphed on a graphing calculator or by hand, and you can visually use the horizontal line test to determine whether they are one-to-one.

a.

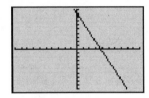

This function passes the horizontal line test because each horizontal line that you could draw across this graph would hit the function only once. Therefore this is a one-to-one function.

b.

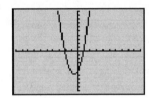

This function fails the horizontal line test because almost any horizontal line that you draw through this graph hits the graph more than once. Therefore this is not a one-to-one function.

c.

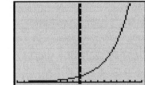

This function passes the horizontal line test because each horizontal line that you could draw across this graph would hit the function only once. Therefore this is a one-to-one function.

d.

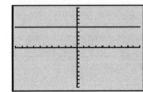

This line is horizontal, so when it is tested by using the horizontal line test, it fails to be one-to-one. Horizontal lines are not one-to-one.

EXAMPLE ③ PRACTICE PROBLEM

Determine whether each of the following functions is a one-to-one function.

a. $g(t) = \dfrac{2}{3}t - 7$ **b.** $h(w) = 0.5w^3 + 3w^2 - 7$

c. $f(x) = -5(0.7)^x$ **d.** $g(x) = \pm\sqrt{x+5}$

Functions that are not one-to-one will not have an inverse function. Before using the steps to find an inverse function we will check if the function has an inverse. If the function does not have an inverse, we will simply say no inverse.

Functions and their inverses have several special relationships. One of these relationships comes from the fact that an inverse function will take an output from the original function and give back the input you originally started with. Ending up with the original input we start with is a result of the inverse function "undoing" what the original function did. When you compose a function and its inverse, the result will always be the input variable. This is best seen with a few examples.

EXAMPLE ❘4❘ COMPOSING INVERSES

Perform the following compositions and simplify.

a. Let $f(x) = 2x + 6$ and $f^{-1}(x) = \dfrac{1}{2}x - 3$. Find $f(f^{-1}(x))$ and $f^{-1}(f(x))$.

b. Let $g(x) = 5x - 20$ and $g^{-1}(x) = \dfrac{1}{5}x + 4$. Find $g(g^{-1}(x))$ and $g^{-1}(g(x))$.

Solution

a. $f(f^{-1}(x)) = 2\left(\dfrac{1}{2}x - 3\right) + 6$

$f(f^{-1}(x)) = x - 6 + 6$

$f(f^{-1}(x)) = x$

$f^{-1}(f(x)) = \dfrac{1}{2}(2x + 6) - 3$

$f^{-1}(f(x)) = x + 3 - 3$

$f^{-1}(f(x)) = x$

b. $g(g^{-1}(x)) = 5\left(\dfrac{1}{5}x + 4\right) - 20$

$g(g^{-1}(x)) = x + 20 - 20$

$g(g^{-1}(x)) = x$

$g^{-1}(g(x)) = \dfrac{1}{5}(5x - 20) + 4$

$g^{-1}(g(x)) = x - 4 + 4$

$g^{-1}(g(x)) = x$

Skill Connection

Recall from Chapter 3 that the composition of functions is the process of making one function the input to the other function.

SC-Example 1:

Let

$f(x) = 2x + 5$

$g(x) = 7x - 9$

Find $f(g(x))$.

$f(g(x)) = 2(7x - 9) + 5$

$f(g(x)) = 14x - 18 + 5$

$f(g(x)) = 14x - 13$

EXAMPLE ④ PRACTICE PROBLEM

Preform the following compositions and simplify.

Let $f(x) = \frac{2}{5}x + 4$ and $f^{-1}(x) = \frac{5}{2}x - 10$. Find $f(f^{-1}(x))$ and $f^{-1}(f(x))$.

The graphs of inverses will also have a simple relationship to the graph of the original function. Because an inverse function basically switches the roles of the input and outputs of the original function, the graph of the inverse will do just that. This results in a graph that is a reflection of the original graph over the line $y = x$. To graph an inverse, you can use techniques you already know, or you can use this reflection property.

EXAMPLE ⑤ GRAPHS OF FUNCTIONS AND THEIR INVERSES

Graph the following functions and their inverses.

a. $f(x) = 2x + 6,$ $\qquad f^{-1}(x) = 0.5x - 3$

b. $h(x) = -\frac{2}{9}x - 6,$ $\qquad h^{-1}(x) = -\frac{9}{2}x - 27$

c. $g(t) = t^3,$ $\qquad g^{-1}(t) = \sqrt[3]{t}$

Solution

a. Because both of these functions are linear, they can be graphed easily. To see how the points on $f(x)$ and $f^{-1}(x)$ are related we will build a small table of points.

$f(x) = 2x + 6$	$f^{-1}(x) = 0.5x - 3$
$(-10, -14)$	$(-14, -10)$
$(-6, -6)$	$(-6, -6)$
$(-3, 0)$	$(0, -3)$
$(0, 6)$	$(6, 0)$
$(5, 16)$	$(16, 5)$

Notice each point for the function has the x and y swapped in the corresponding point for the inverse function.

Putting both graphs on the same axis, and showing the line $y = x$, we get the following.

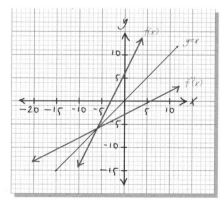

b. Again these functions are both linear, so we can graph them easily.

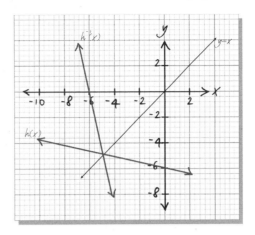

c. These functions are harder to graph by hand. We will plot several points for the original function and use the reflection properties of inverses to get a graph of the inverse. Therefore, for each point that we find for the original function $g(x) = x^3$ we will reverse the x and y values to find a point on the inverse function $g^{-1}(x) = \sqrt[3]{x}$.

$g(x) = x^3$	$g^{-1}(x) = \sqrt[3]{x}$
$(-1.4, -2.744)$	$(-2.744, -1.4)$
$(-1, -1)$	$(-1, -1)$
$(-0.4, -0.064)$	$(-0.064, -0.4)$
$(0, 0)$	$(0, 0)$
$(0.2, 0.008)$	$(0.008, 0.2)$
$(0.6, 0.216)$	$(0.216, 0.6)$
$(1, 1)$	$(1, 1)$
$(1.2, 1.728)$	$(1.728, 1.2)$

Plotting the points for the function and the inverse we get the following.

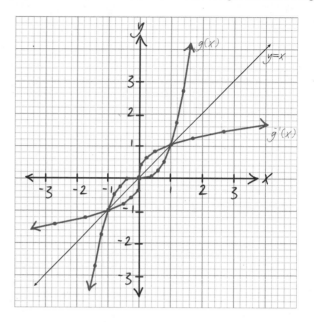

In graphing these on a graphing calculator, it is best to use the ZOOM Square feature so that the graph will not be distorted by the shape of the calculator screen.

6.1 Exercises

For Exercises 1 through 10, use the horizontal line test to determine whether or not the function is one-to-one.

1.

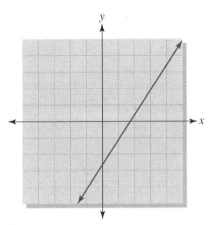

2.

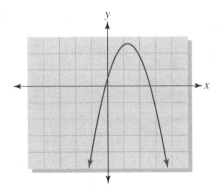

3.

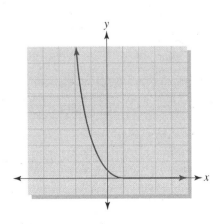

4.

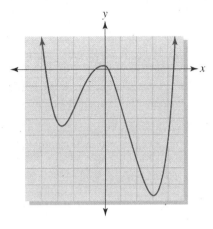

5. $f(x) = \dfrac{1}{2}x + 3$

6. $g(x) = -\dfrac{2}{3}x + 7$

7. $h(x) = 2.5x^2 + 3x - 9$

8. $f(x) = -0.4(x - 3.5)^2 + 8$

9. $g(x) = 3(1.2)^x$

10. $h(x) = -25(0.7)^x$

11. The number of homicides, N, of 15- to 19-year-olds in the United States t years after 1990 can be represented by the equation

$$N(t) = -315.9t + 4809.8$$

Source: Based on data from Statistical Abstract 2001.

    a. Find the inverse for this model.

    b. If the model has a domain of $[-1, 12]$ and a range of $[1019, 5125.7]$, give the domain and range for the inverse.

    c. Estimate the year in which there were 2000 homicides of 15- to 19-year-olds.

12. $P(b) = 5.5b - 2500$ represents the profit in dollars from selling b books.

    a. Given a domain for $P(b)$ of $0 \le b \le 5000$, find the range for this model.

    b. Find an inverse for this model.

c. Estimate the profit from selling 1000 books.

d. Estimate the number of books you would need to sell to make $5000 profit.

e. Give the domain and range for the inverse.

13. The population of the United States during the 1990s can be estimated by the equation

$$P(t) = 2.57t + 249.78$$

where P is the population in millions t years since 1990.

Source: Based on data from Statistical Abstract 2001.

a. Find an inverse for this model.

b. Estimate the year when the U.S. population reached 260 million people.

c. What are the input and output variables for the inverse?

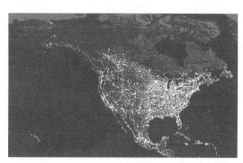

Population density of the United States

14. The population of Colorado in millions t years since 1990 can be modeled by

$$P(t) = 0.0875t + 3.285$$

a. Find an inverse for this model.

b. Estimate when Colorado's population reached 3.5 million people.

c. Give a reasonable domain and range for the inverse.

15. The average number of pounds of fruits and vegetables per person that each American eats can be modeled by

$$T(y) = 6.056y + 601.39$$

pounds per person y years since 1980.

a. Find an inverse for this model.

b. Estimate when the average number of pounds of fruits and vegetables per person reached 650.

c. Give a reasonable domain and range for this inverse.

16. The amount of money Hip Hop Math Records will pay "Math Dude" can be modeled by
$$M(c) = 10000 + 1.5c$$

where $M(c)$ is the money in dollars paid to "Math Dude" when c CDs are sold.

a. Find an inverse for this model.

b. Find how many CDs must be sold for "Math Dude" to earn $40,000.

c. If the record company believes they can sell between 10,000 and 50,000 CDs find a domain and range for the inverse function.

For Exercises 17 through 24, find the inverse of each function

17. $f(x) = 3x + 5$

18. $f(x) = 5x + 7$

19. $g(t) = -4t + 8$

20. $h(x) = -6x - 10$

21. $h(x) = \dfrac{2}{3}x - 9$

22. $f(x) = -\dfrac{5}{7}x + 4$

23. $P(t) = -2.5t - 7.5$

24. $W(x) = 2.4x + 3.7$

For Exercises 25 through 30, compose the two given functions and determine whether they are inverses.

25. $f(x) = 3x - 9$ and $g(x) = \dfrac{1}{3}x + 3$

26. $f(x) = 5x - 30$ and $g(x) = \dfrac{1}{5}x + 6$

27. $f(x) = 4x + 12$ and $g(x) = 0.25x - 3$

28. $h(x) = 0.6x + 5$ and $g(x) = 0.4x - 2$

29. $f(x) = \frac{1}{7}x + 21$ and $h(x) = 7x + 3$

30. $g(x) = -8x - 12$ and $h(x) = -\frac{1}{8}x - 1\frac{1}{2}$

For Exercises 31 through 36, graph the given functions and their inverses.

31. $f(x) = 3x - 9$

32. $f(x) = 4x + 12$

33. $g(t) = -4t + 8$

34. $f(x) = -\frac{5}{7}x + 4$

35. $f(x) = 2^x$ (Hint: build a table of points and reverse x and y to find points on the inverse.)

36. $g(x) = 100(0.25)^x$ (Hint: build a table of points and reverse x and y to find points on the inverse.)

Introduction to Logarithms

DEFINITION OF LOGARITHMS

As we can see from the horizontal line test, most linear functions have inverses, but quadratic functions do not. The exponential function also passes the horizontal line test, so it also must have an inverse function.

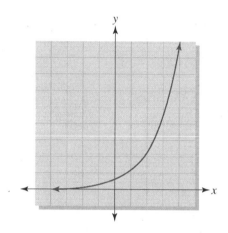

When we try to find the inverse function for an exponential, trying to solve for the input variable leads us to a basic problem. There is no way, using the arithmetic operations we know, to get a variable out of an exponent. This problem leads us to a new type of function that at the most basic level can be defined as the inverse function of an exponential. We use the symbol $\log_b$, read "log base b," and call this new function a **logarithm.** A logarithm at its core asks the question "What exponent of the base b gives us what's inside?"

DEFINITIONS

Logarithmic function: An inverse for an exponential function.
A basic logarithmic function can be written in the form

$$f(x) = \log_b x \quad \text{read log base } b \text{ of } x.$$

where b is a real number greater than zero and not equal to one. By definition logarithms and exponentials have an inverse relationship.

$$f(x) = y = b^x \qquad f^{-1}(x) = y = \log_b x$$

Base: The constant b is called the base of the logarithm function. If no base is given, then that log is assumed to have base 10.

A logarithm asks the basic question: "What exponent of the base b gives us what's inside the log?"

EXAMPLE   EVALUATING BASIC LOGARITHMS

Evaluate the following logarithms.

a. $\log_5 25$ **b.** $\log_2 16$ **c.** $\log 1000$

Solution

a.   With each of these logarithms you will want to ask yourself the basic question "What exponent of 5 will give us 25?" In this case we know that $5^2 = 25$, so we know an exponent of 2 will give us 25. Therefore we have $\log_5 25 = 2$.

b.   Ask, "what exponent of 2 will give me 16?" We should know that $2^4 = 16$, so an exponent of 4 will give you 16. Therefore we have that $\log_2 16 = 4$.

c.   In this case no base is written in the logarithmic expression. This implies that this is a common logarithm and always has the base 10. Therefore the question becomes "What exponent of 10 will give me 1000?" Because $10^3 = 1000$, we know that $\log 1000 = 3$.

EXAMPLE   1   PRACTICE PROBLEM

Evaluate the following logarithms.

a. $\log 10000$ **b.** $\log_7 49$

BASIC RULES FOR LOGARITHMS

As you can see from example 1, evaluating logarithms is pretty simple if you can recognize the exponent that you need. Some logarithms have the same value for any base. In particular, in taking the logarithm of 1, no matter what the base is, the logarithm will always equal zero. This follows from the exponent rule that says anything, except zero, to the zero power will be one.

$$b^0 = 1 \quad \text{so} \quad \log_b 1 = 0$$

Another basic rule says that if you take the logarithm of the base to a power, you will get that power back. Using the basic question a logarithm asks for $\log_4 4^9$ we get "what exponent of 4 will give me 4^9?" an exponent of 9 is the only exponent that would make sense. These rules are all related directly to the definition of a logarithm and make up some of the most basic rules for logarithms.

BASIC LOGARITHM RULES

$$\log_b 1 = 0 \qquad \text{since } b^0 = 1$$

The logarithm of 1 for any base logarithm will always equal zero.

$$\log_b b = 1 \qquad \text{since } b^1 = b$$

The logarithm of its base is always equal to 1.

$$\log_b(b^m) = m \qquad \text{since } b^m = b^m$$

The logarithm of its base to a power is just that power.

EXAMPLE 2 EVALUATING LOGARITHMS USING THE BASIC RULES

Evaluate the following logarithms using the basic logarithm rules.

a. $\log 10^5$ b. $\log_3 1$ c. $\log_9 9$

Solution

a. The base of this logarithm is understood to be 10. We have the log of its base raised to a power so the logarithm will be equal to the power.

$\log 10^5 = 5$

b. We are taking the log of 1 so the log equals zero.

c. This is a log of its own base so the log equals 1.

EXAMPLE 2 PRACTICE PROBLEM

Evaluate the following logarithms using the basic logarithm rules.

a. $\log_7 7^8$ b. $\log_4 1$

Although logarithms can be written with any positive base other than 1, there are two logarithms used the most. The "common" logarithm is base 10 and is written as "log" without a base present. This logarithm is used often in science when very large or very small numbers are needed. One other base that is often used in science and business is base e. When a logarithm has a base of e the logarithm is written as "Ln" and a base of e is assumed. A logarithm with base e is most often called the **natural logarithm.** One of the nicest features of calculators is that the common and natural logarithms can be found with the touch of a button.

EXAMPLE 3 EVALUATING LOGARITHMS USING A CALCULATOR

Use your calculator to evaluate the following logarithms.

a. $\log 500$ b. $\ln 13.4$

Solution

Be sure to use the correct button on your calculator.

```
log(500)
        2.698970004
ln(13.4)
        2.595254707
```

a. $\log 500 \approx 2.699$

b. $\ln 13.4 \approx 2.595$

CHANGE OF BASE FORMULA

If you want to calculate a logarithm with a base other than 10 or e, you can start by trial and error, using different exponents of the base, or you will need to change the base of the logarithm to use a common or natural logarithm calculation on your calculator.

Using Your TI Graphing Calculator

To evaluate logarithms on your calculator, you use the LOG or LN button on the left side of the calculator.

These buttons will calculate only common or natural logarithms. Be sure to use the one with the base that you want.

When you use these buttons, they will automatically start the expression with a left parenthesis. It is a good idea to end with a right parenthesis so that you get the values you want. This will be more important when you get to later sections and are doing more complicated calculations.

CONCEPT INVESTIGATION I WHAT DO I DO WITH THAT BASE?

Since we have both log base 10 and log base e buttons on our calculator, but no other bases, we need a way to calculate logarithms of bases other than 10 or e. Answer the following questions to help discover the way to change a base of a logarithm.

a. We will start with $\log_5 25$. We know that this logarithm asks "what exponent of 5 will give me 25?" We know that 5 squared gives us 25 so $\log_5 25 = 2$. Now we will try a few other calculations with log base 10 to see if we can find this same answer on our calculator.

$\log 25 + \log 5$	$\log 25 - \log 5$
2.097	
$\log 25 \cdot \log 5$	$\dfrac{\log 25}{\log 5}$

Which of these calculations were equal to 2?

b. Now let's consider $\log_3 81$. We know that this logarithm asks "what exponent of 3 will give me 81?" We know that 3 raised to the 4th power will give us 81 so $\log_3 81 = 4$. Now we will try a few other calculations with ln to see if we can find this same answer on our calculator.

$\ln 81 + \ln 4$	$\ln 81 - \ln 4$
3.008	
$\ln 81 \cdot \ln 4$	$\dfrac{\ln 81}{\ln 4}$

Which of these calculations were equal to 4?

c. Now let's try one we don't know the answer to, $\log_7 250$. Since we do not know what this logarithm equals we will need to do each calculation and check it by raising the base to that power to see if we get 250. Either log or ln will work here so we will use ln.

```
ln(250)+ln(7)
        7.467371067
7^Ans
        2044846.219
```

$\ln 250 + \ln 7$ **7.467** **Check this answer.** $7^{7.467} \approx 2044846.2$	$\ln 250 - \ln 7$
$\ln 250 \cdot \ln 7$	$\dfrac{\ln 250}{\ln 7}$

Which calculation gave us the result we were looking for?

From the concept investigation we can see that using division we can change from one base to a more convenient base. The order that you divide is important so be careful to take the log of the inside and divide by the log of the old base. The change of base formula is a simple way to change a logarithm from one base to another.

CHANGE OF BASE FORMULA FOR LOGARITHMS

$$\log_b x = \frac{\log_c x}{\log_c b}$$

To change the base of a logarithm from base b to another base c, take the $\log_c$ of the inside and divide by the $\log_c$ of the old base.

Most often, we change to base 10 or base e because our calculators can evaluate these bases. Therefore, the change of base formula most often looks like

$$\log_b x = \frac{\log x}{\log b} \qquad \text{or} \qquad \log_b x = \frac{\ln x}{\ln b}$$

EXAMPLE 4 USING THE CHANGE OF BASE FORMULA

Evaluate the following logarithms.

a. $\log_5 114$ **b.** $\log_2 0.365$

Solution

Because these logarithms are not of base 10 or e, we will need to change their base to use our calculator to evaluate them. You can change the base to either 10 or e for either of these examples. We change one to base 10 and the other to base e.

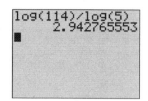

a. Using a base 10 logarithm, we get

$$\log_5 114$$

$$\log_5 114 = \frac{\log 114}{\log 5}$$ Use the change of base formula.

$$\log_5 114 \approx 2.943$$

b. Using a natural logarithm, we get

$$\log_2 0.365$$

$$\log_2 0.365 = \frac{\ln 0.365}{\ln 2}$$ Use the change of base formula.

$$\log_2 0.365 \approx -1.454$$

EXAMPLE ④ PRACTICE PROBLEM

Evaluate the following logarithms on your calculator.

a. $\log_3 278$ **b.** $\log_{17} 11$

The goal of this section is that you start to understand the basic concept of a logarithm and how it relates to exponential functions. One of the most important skills that you need to learn is to rewrite a problem given in exponential form into logarithm form and vice versa. When working with logarithms and exponentials, remember that the bases are related and that the exponent in an exponential is the same as the result from the logarithm.

LOGARITHM AND EXPONENTIAL FORMS

$$\log_b n = m$$
$$n = b^m$$

The base of the logarithm is the same as the base of the related exponential function. The result from the logarithm is the exponent in the exponential function.

When rewriting a logarithm, you can think, "The base raised to the outside equals the inside."

$$\log_5 125 = 3$$
$$125 = 5^3$$

The domain and range of exponential and logarithmic functions are related because these functions are inverses of one another.

$$f(x) = b^x \qquad f^{-1}(x) = \log_b x$$

Domain: $(-\infty, \infty)$ Domain: $(0, \infty)$
Range: $(0, \infty)$ Range: $(-\infty, \infty)$

EXAMPLE **5** REWRITING IN LOGARITHM OR EXPONENTIAL FORM

Rewrite each exponential equation into logarithm form.

a. $7^5 = 16807$

b. $4.5^3 = 91.125$

Rewrite each logarithm equation into exponential form.

c. $\log_3 243 = 5$

d. $\log_{2.5} 15.625 = 3$

Solution

a. The base is 7 so the base of the logarithm will also be 7. The exponent is 5 so the logarithm will equal 5.

$$\log_7 16807 = 5$$

b. The base is 4.5 so the base of the logarithm will also be 4.5. The exponent is 3 so the logarithm will equal 3.

$$\log_{4.5} 91.125 = 3$$

c. The base is 3 so the base of the exponential will also be 3. The logarithm equals 5 so the exponent will equal 5.

$$243 = 3^5$$

d. The base is 2.5 so the base of the exponential will also be 2.5. The logarithm equals 3 so the exponent will equal 3.

$$15.625 = 2.5^3$$

INVERSES

Rewriting exponential functions into logarithmic form and logarithm functions into exponential form is actually writing their inverse functions.

LOGARITHMS AND EXPONENTIALS AS INVERSES

$$f(x) = b^x$$
$$f^{-1}(x) = \log_b x$$

or

$$f(x) = \log_b x$$
$$f^{-1}(x) = b^x$$

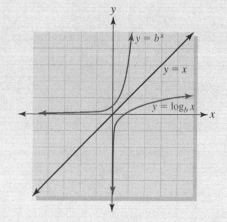

EXAMPLE   6   INVERSES FOR LOGARITHMS AND EXPONENTIALS

Find the inverses for the following functions.

a. $f(x) = 3^x$ **b.** $g(t) = \log t$

c. $h(x) = 5(4)^x$ **d.** $f(x) = e^x$

Solution

We basically need to follow the same steps as finding an inverse in a problem without a context that we learned in Section 6.1. But when we solve, we will rewrite exponentials into logarithmic form and logarithms into exponential form to isolate the variable we want.

a.   Since this function is an exponential the inverse will be a logarithm.

$f(x) = 3^x$

$y = 3^x$	Take the function out of function notation.
$\log_3 y = x$	Solve for x by writing it in logarithm form.
$\dfrac{\log y}{\log 3} = x$	Rewrite using the change of base formula.
$\dfrac{\log x}{\log 3} = y$	Interchange the variables.
$f^{-1}(x) = \dfrac{\log x}{\log 3}$	Write in function notation.

b.   Since this function is a logarithm the inverse will be an exponential.

$g(t) = \log t$

$y = \log t$	Take the function out of function notation.
$10^y = t$	Solve for t by writing it in exponential form.
$10^t = y$	Interchange the variables.
$g^{-1}(t) = 10^t$	Write in function notation.

c.   Since this function is an exponential the inverse will be a logarithm.

$h(x) = 5(4)^x$

$y = 5(4)^x$	Take the function out of function notation.
$\dfrac{y}{5} = 4^x$	Isolate the exponential base.
$\log_4\left(\dfrac{y}{5}\right) = x$	Solve for x by writing in exponential form.
$\dfrac{\log\left(\dfrac{y}{5}\right)}{\log 4} = x$	Rewrite using the change of base formula.
$\dfrac{\log\left(\dfrac{x}{5}\right)}{\log 4} = y$	Interchange the variables.
$h^{-1}(x) = \dfrac{\log\left(\dfrac{x}{5}\right)}{\log 4}$	Write in function notation.

d. Since the base of the exponential is e, the inverse will be the natural logarithm.
$$f(x) = e^x$$
$$f^{-1}(x) = \ln x$$

SOLVING SIMPLE LOGARITHMIC EQUATIONS

Simple logarithm equations can be solved by rewriting the logarithm into exponential form.

EXAMPLE **7** SOLVING SIMPLE LOGARITHM EQUATIONS

Solve each logarithm equation by rewriting it in exponential form.

a. $\log x = 2$

b. $\log_4 x = 5$

Solution

a. $\log x = 2$

$\qquad 10^2 = x \qquad$ **Rewrite as an exponential and calculate.**

$\qquad x = 100$

b. $\log_4 x = 5$

$\qquad 4^5 = x \qquad$ **Rewrite as an exponential and calculate.**

$\qquad x = 1024$

EXAMPLE **7** **PRACTICE PROBLEM**

Solve each logarithm equation by rewriting it in exponential form.

a. $\ln x = 4$

b. $\log_2 (5x) = 6$

6.2 Exercises

For Exercises 1 through 12, evaluate each logarithm without a calculator.

1. $\log 1000$

2. $\log_2 32$

3. $\log_5 125$

4. $\log_4 16$

5. $\log_7 1$

6. $\ln 1$

7. $\ln(e^3)$

8. $\log_3(3^7)$

9. $\log_{19} 19$

10. $\log_6 6$

11. $\log_2 256$

12. $\log_3 243$

13. $\log_5\left(\dfrac{1}{5}\right)$

14. $\log_2\left(\dfrac{1}{16}\right)$

15. $\log 0.1$

16. $\log 0.0001$

17. $\log_7\left(\dfrac{1}{49}\right)$

18. $\log_3\left(\dfrac{1}{27}\right)$

For Exercises 19 through 28, evaluate the following logarithms with a calculator.

19. $\log 125$

20. $\log 3000$

21. $\ln 45$

22. $\ln 543$

23. $\log_7 25$

24. $\log_5 43.2$

25. $\log_2 0.473$

26. $\log_9 0.68$

27. $\log_{14} 478$

28. $\log_{25} 5860$

For Exercises 29 through 34, rewrite the following logarithm equations into exponential form.

29. $\log 1000 = 3$

30. $\ln(e^5) = 5$

31. $\log_2(8) = 3$

32. $\log 0.01 = -2$

33. $\log_5\left(\dfrac{1}{25}\right) = -2$

34. $\log_4\left(\dfrac{1}{64}\right) = -3$

For Exercises 35 through 40, rewrite each exponential equation into logarithm form.

35. $2^{10} = 1024$

36. $3^5 = 243$

37. $3^{2x} = 729$

38. $b^y = c$

39. $\left(\dfrac{1}{2}\right)^3 = \dfrac{1}{8}$

40. $\left(\dfrac{1}{3}\right)^4 = \dfrac{1}{81}$

For Exercises 41 through 50, find the inverse of each function.

41. $f(x) = 7^x$

42. $B(t) = e^t$

43. $h(c) = 10^c$

44. $g(t) = 2^t$

45. $m(r) = 5^r$

46. $f(x) = 12^x$

47. $f(x) = 3(4)^x$

48. $g(x) = 6(5)^x$

49. $h(x) = \dfrac{1}{2}(9)^x$

50. $f(t) = \dfrac{1}{3}(2)^t$

For Exercises 51 through 60, solve each logarithm equation by rewriting it in exponential form.

51. $\log x = 3$

52. $\log x = 5$

53. $\ln t = 2$

54. $\ln t = 4$

55. $\log_3 w = 2.4$

56. $\log_9 x = 0.5$

57. $\log(8t) = 2$

58. $\log(2x) = 0.5$

59. $\ln(3x) = 4$

60. $\log_7(4x) = 3$

Graphing Logarithmic Functions

GRAPHING LOGARITHMIC FUNCTIONS

In Section 6.2 we learned that logarithmic functions are defined as inverses of exponential functions. This inverse relationship helps us to investigate the graph of logarithm functions using the information we know about the related exponential graphs. Recall from Section 6.1 that the graph of an inverse function is a reflection of the graph of the original function over the line $y = x$. With this relationship in mind lets look at some exponential graphs and their related logarithm graphs.

 CONCEPT INVESTIGATION I HOW DO WE BUILD A
LOGARITHM GRAPH?.........................

a. Fill in the following table of values for the function $f(x) = 2^x$

x	$f(x) = 2^x$
-2	
-1	
0	
1	
2	
3	
4	
5	
6	
7	
8	

b. What kind of scale should we use on the x-axis for this graph?

      0.5   1   2   5   10   other:_____

c. What kind of scale should we use on the y-axis for this graph?

      0.5   1   2   5   10   other:_____

d. Sketch the graph of the function in part a.

e. Use the table you completed in part a to complete the table for the inverse function. (Remember that the x and y values will be swapped.)

x	$f^{-1}(x) = \log_2 x = \dfrac{\log x}{\log 2}$

f. What kind of scale should we use on the x-axis for this graph?

        0.5   1   2   5   10   other:_____

g. What kind of scale should we use on the y-axis for this graph?

        0.5   1   2   5   10   other:_____

h. Are these scales related to the scales you chose in parts b and c? If so, how?

i. Sketch the graph of the function in part e.

From the concept investigation, we can see that using the related exponential graph to sketch the graph of a logarithm is a reasonable approach. If we look at the characteristics of an exponential graph, we can find the basic characteristics of a logarithmic function's graph.

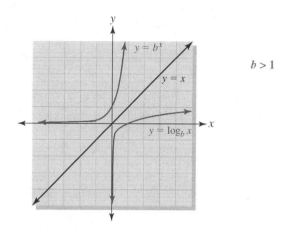

When the base is greater than 1 the graphs have the following characteristics.

Characteristics of Exponential Graph $b > 1$	Characteristics of Logarithmic Graphs $b > 1$
No x-intercept.	No y-intercept.
y-intercept $(0,1)$.	x-intercept $(1,0)$.
The x-axis is a horizontal asymptote.	The y-axis is a vertical asymptote.
Graph increases slowly and then rapidly.	Graph increases rapidly and then slowly.

The growth of a logarithmic function gets slower and slower as the input values increase. If you consider the common logarithm, log x, it takes a large value of x to get a relatively small y value.

x	1	10	100	1,000	10,000
$f(x) = \log x$	0	1	2	3	4

As you can see from this table to get an output of 3 the input must be 1,000. To get an output of 4 you need the input to be 10,000. The very slow growth of a logarithm function makes graphing these functions harder. Looking at the graph of log x on a standard window and even an expanded window shows how slowly the graph grows as x increases.

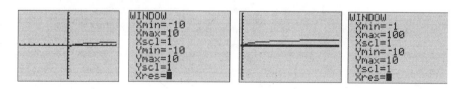

EXAMPLE ■ GRAPH A LOGARITHMIC FUNCTION

Sketch the graph of $f(x) = \log_3 x$.

Solution

Because this logarithm has a base of 3, the inverse will be the exponential function
$$f^{-1}(x) = 3^x$$
We will use this exponential function to create a table of points to plot for the graph.

Points for a graph of the inverse.   Points for the logarithm graph.

x	$f^{-1}(x) = 3^x$
−2	$\frac{1}{9}$
−1	$\frac{1}{3}$
0	1
1	3
2	9
3	27
4	81
5	243

x	$f(x) = \log_3 x$
$\frac{1}{9}$	−2
$\frac{1}{3}$	−1
1	0
3	1
9	2
27	3
81	4
243	5

To plot the logarithm graph we need a large scale for the x-axis and a small scale for the y-axis.

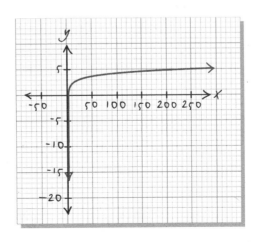

EXAMPLE ① PRACTICE PROBLEM

Sketch the graph of $f(x) = \log_5 x$.

Now we will consider the graphs of logarithms with bases between 0 and 1.

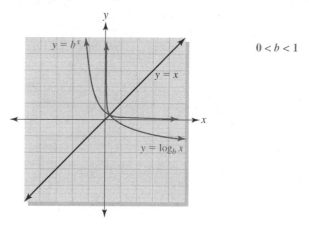

$0 < b < 1$

When the base is between 0 and 1 the graphs have the following characteristics.

Characteristics of Exponential Graph $0 < b < 1$	Characteristics of Logarithmic Graphs $0 < b < 1$
No x-intercept.	No y-intercept.
y-intercept $(0,1)$.	x-intercept $(1,0)$.
The x-axis is a horizontal asymptote.	The y-axis is a vertical asymptote.
Graph decreases rapidly and then slowly.	Graph decreases rapidly and then slowly.

EXAMPLE   2   GRAPH A LOGARITHMIC FUNCTION

Sketch the graph of $f(x) = \log_{0.5} x$.

Solution

Because this logarithm has a base of 0.5, the inverse will be the exponential function
$$f^{-1}(x) = (0.5)^x$$

We will use this exponential function to create a table of points to plot for the graph.

Points for a graph of the inverse.   Points for the logarithm graph.

x	$f^{-1}(x) = (0.5)^x$
−10	1024
−8	256
−6	64
−4	16
−2	4
0	1
2	0.25
4	0.0625

x	$f(x) = \log_{0.5} x$
1024	−10
256	−8
64	−6
16	−4
4	−2
1	0
0.25	2
0.0625	4

To plot the logarithm graph we need a large scale for the *x*-axis and a small scale for the *y*-axis.

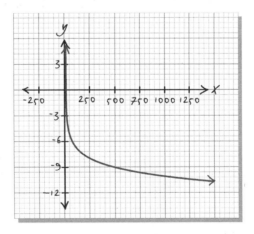

EXAMPLE ② PRACTICE PROBLEM

Sketch the graph of $f(x) = \log_{0.25} x$.

We can see from both of these examples that basic logarithm graphs have a few things in common. They have the *y*-axis as a vertical asymptote, they have an *x*-intercept of (1,0), and they grow or decay rapidly at first and then slowly as the inputs increase.

DOMAIN AND RANGE OF LOGARITHMIC FUNCTIONS

There are several ways to think about the domain and range of a logarithmic function. The inverse relationship between logarithmic and an exponential functions can be used to relate the domain and range of exponential functions to the domain and range of logarithmic functions. We can also consider the basic graphs of logarithmic functions to find their domain and range.

In Chapter 5 we found that the domain and range of an exponential function were all real numbers, $(-\infty, \infty)$, and all positive real numbers, $(0, \infty)$, respectively. Since logarithms are the inverse of exponentials the domain and range are simply switched. Therefore the basic domain for a logarithm function will be all positive real numbers or $(0, \infty)$, and the range will be all real numbers or $(-\infty, \infty)$.

We can confirm this domain and range using the basic graphs of logarithms.

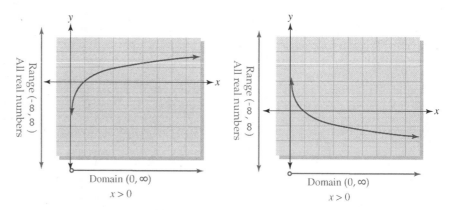

The domain is restricted to only the positive real numbers because you cannot take the logarithm of zero or a negative number

EXAMPLE **3** DOMAIN AND RANGE OF A LOGARITHMIC FUNCTION

Give the domain and range of $f(x) = \log_7 x$.

Solution

Because this is a basic logarithm function we know
   Domain: All positive real numbers, or $(0, \infty)$
   Range: All real numbers, or $(-\infty, \infty)$

6.3 Exercises

For Exercises 1 through 4, use the given graphs to estimate the answers.

1. Given the graph of $f(x)$

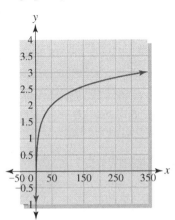

  a. Estimate $f(50)$.

  b. Estimate $f(x) = 2.5$.

2. Given the graph of $g(x)$

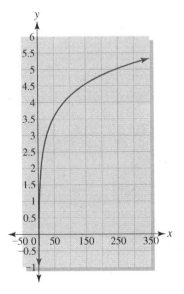

  a. Estimate $g(150)$.

  b. Estimate $g(x) = 5$.

3. Given the graph of $h(x)$

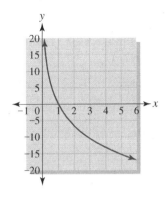

  a. Estimate $h(3)$.

  b. Estimate $h(x) = -15$.

4. Given the graph of $f(x)$

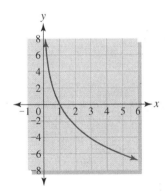

  a. Estimate $f(5)$.

  b. Estimate $f(x) = -4$.

For Exercises 5 through 18, sketch the graph of the following functions.

5. $f(x) = \log x$

6. $f(x) = \ln x$

7. $g(x) = \log_4 x$

8. $h(x) = \log_6 x$

9. $f(x) = \log_9 x$

10. $g(x) = \log_8 x$

11. $f(x) = \log_{20} x$

12. $h(x) = \log_{25} x$

13. $g(x) = \log_{0.2} x$

14. $f(x) = \log_{0.4} x$

15. $h(x) = \log_{0.9} x$

16. $g(x) = \log_{0.8} x$

17. $h(x) = \log_{0.6} x$

18. $f(x) = \log_{0.1} x$

19. Give the domain and range of the function graphed in exercise 1.

20. Give the domain and range of the function graphed in exercise 2.

21. Give the domain and range of the function graphed in exercise 3.

22. Give the domain and range of the function graphed in exercise 4.

23. Give the domain and range of the function given in exercise 5.

24. Give the domain and range of the function given in exercise 6.

25. Give the domain and range of the function given in exercise 7.

26. Give the domain and range of the function given in exercise 8.

27. Give the domain and range of the function given in exercise 13.

28. Give the domain and range of the function given in exercise 14.

29. Give the domain and range of the function given in exercise 15.

30. Give the domain and range of the function given in exercise 16.

Properties of Logarithms

PROPERTIES OF LOGARITHMS

This section will cover the basic properties of logarithms that we will need in the rest of this chapter to solve exponential and logarithm problems.

Because logarithms are inverses of exponential functions, the basic rules for exponents that we learned in Section 3.1 will have their related rules for logarithms. Use the concept investigation to find the related properties of logarithms.

 CONCEPT INVESTIGATION ∣ **ARE THESE LOGS THE SAME?** ....

Evaluate the following logarithm expressions on your calculator. Compare the results for the expression in the left column to the corresponding expression in the right column.

1a. $\ln(7 \cdot 2) =$ **1b.** $\ln 7 + \ln 2 =$

2a. $\log(8 \cdot 3) =$ **2b.** $\log 8 + \log 3 =$

3a. $\ln(5 \cdot 11) =$ **3b.** $\ln 5 + \ln 11 =$

4a. $\log 100 =$ **4b.** $\log 10 + \log 10 =$

Describe in your own words the relationship between these two columns of logarithm expressions.

b. Evaluate the following logarithm expressions on your calculator. Compare the results for the expression in the left column to the corresponding expression in the right column.

1a. $\log\left(\dfrac{30}{6}\right) =$ **1b.** $\log 30 - \log 6 =$

2a. $\ln\left(\dfrac{45}{5}\right) =$ **2b.** $\ln 45 - \ln 5 =$

3a. $\ln\left(\dfrac{5000}{45}\right) =$ **3b.** $\ln 5000 - \ln 45 =$

4a. $\log\left(\dfrac{1000}{10}\right) =$ **4b.** $\log 1000 - \log 10 =$

Describe in your own words the relationship between these two columns of logarithm expressions.

These two relationships are stated in the **product rule for logarithms** and the **quotient rule for logarithms.**

THE PRODUCT RULE FOR LOGARITHMS

$$\log_b(mn) = \log_b m + \log_b n$$

A logarithm of any base with multiplication inside can be written as two separate logarithms added together.

$$\log_b(5 \cdot 9) = \log_b 5 + \log_b 9$$

THE QUOTIENT RULE FOR LOGARITHMS

$$\log_b\left(\frac{m}{n}\right) = \log_b m - \log_b n$$

A logarithm of any base with division inside can be written as two separate logarithms subtracted from one another.

$$\log_b\left(\frac{50}{8}\right) = \log_b 50 - \log_b 8$$

These two rules allow us to either take apart a complex logarithm or put two logarithms together into one. In different circumstances we want to do either of these two processes. Most often, though, we will be putting two logarithms together to become one logarithm so that we can solve a simpler equation.

EXAMPLE   **1** EXPANDING AND COMBINING LOGARITHMS

a. Expand the following logarithms as separate simpler logarithms.

i. $\ln(5 \cdot 17)$

ii. $\log\left(\frac{35}{8}\right)$

iii. $\log_4(13xy)$

iv. $\log_5\left(\frac{mn}{9}\right)$

b. Write the following logarithms as a single logarithm.

i. $\log 8 + \log 7$

ii. $\ln(x+2) + \ln(x+5)$

iii. $\ln(3x) - \ln(5z)$

iv. $\log_7(2m) - \log_7 n$

Solution

a. To separate these logarithms into simpler logarithms, we can use the product and quotient rules for logarithms.

i. $\ln 5 + \ln 17$ Make separate logarithms separated by addition.

ii. $\log 35 - \log 8$ Make separate logarithms separated by subtraction.

iii. $\log_4 13 + \log_4 x + \log_4 y$

iv. $\log_5 m + \log_5 n - \log_5 9$

b. Combining these logarithms will also use the product and quotient rules for logarithms.

   i. $\log 56$ **Multiply the insides of the logarithms.**

   ii. $\ln[(x+2)(x+5)] = \ln(x^2 + 7x + 10)$ **Multiply the insides of the logarithms using FOIL.**

   iii. $\ln\left(\dfrac{3x}{5z}\right)$ **Divide the insides of the logarithms.**

   iv. $\log_7\left(\dfrac{2m}{n}\right)$ **Divide the insides of the logarithms. The base stays the same.**

EXAMPLE PRACTICE PROBLEM

a. Expand the following logarithms as separate simpler logarithms.

   i. $\ln(5ab)$ **ii.** $\log\left(\dfrac{7n}{m}\right)$

b. Write the following logarithms as a single logarithm.

   i. $\ln(18s) - \ln(3t)$ **ii.** $\log_7 5 + \log_7 m + \log_7(3n)$

   iii. $\log(x+3) + \log(x-7)$

 CONCEPT INVESTIGATION 2 WHAT HAPPENS TO THE POWER? ·······················

Evaluate the following logarithm expressions on your calculator. Compare the results for the expression in the left column to the expression in the right column.

1a. $\log(7^3) =$ **1b.** $3\log 7 =$

2a. $\ln(8^5) =$ **2b.** $5\ln 8 =$

3a. $\log(7^{-2}) =$ **3b.** $-2\log 7 =$

Describe in your own words the relationship between these two columns of logarithm expressions.

··

    This relationship is one of the most useful rules for logarithms and will be used in a very important role in most exponential and logarithm equations. This rule is stated in the **power rule for logarithms.**

THE POWER RULE FOR LOGARITHMS

$$\log_b(m^n) = n \log_b m$$

A logarithm of any base with a power inside can be written as that logarithm with that power now being multiplied to the front.

$$\log_b(5^4) = 4 \log_b 5$$

Using the power rule for exponents together with the product and quotient rules allows you to combine or simplify logarithms of all kinds.

EXAMPLE   **2**   **EXPANDING AND COMBINING LOGARITHMS WITH EXPONENTS**

a. Expand the following logarithms as separate simpler logarithms without any exponents.

   i. $\ln(5x^2)$

   ii. $\log\left(\dfrac{2x^3}{5}\right)$

   iii. $\log_4(9x^3y^2)$

   iv. $\log_5\left(\dfrac{m^2n^5}{2z^3}\right)$

   v. $\log(\sqrt{5}xy)$

b. Write the following logarithms as a single logarithm with a coefficient of one.

   i. $\log 8 + 3 \log x + 4 \log y$ **ii.** $\ln 3 + 2 \ln 5a - \ln b$

   iii. $\ln x - 5 \ln z + 2 \ln y$

   iv. $\log_7 2 + \log_7 m - 3 \log_7 n + \log_7 5$

Solution

a. To separate these logarithms into simpler logarithms, we can use the product and quotient rules for logarithms.

   i.

   $\ln 5x^2$

   $\ln 5 + \ln x^2$ Make into separate logs with addition.

   $\ln 5 + 2 \ln x$ Use the power rule for logarithms to bring the exponent down.

   ii. $\log 2 + 3 \log x - \log 5$

   iii. $\log_4 9 + 3 \log_4 x + 2 \log_4 y$

   iv.

   $$\log_5\left(\dfrac{m^2n^5}{2z^3}\right)$$ Expand using the product and quotient rules.

   $$\log_5(m^2) + \log_5(n^5) - \log_5 2 - \log_5(z^3)$$ Bring the exponents down using the power rule.

   $$2 \log_{(5)} m + 5 \log_5 n - \log_5 2 - 3 \log_5 z$$

v.

$$\log\left((5xy)^{\frac{1}{2}}\right)$$

Change the square root into a fraction exponent.

$$\frac{1}{2}\log(5xy)$$

Use the power rule to bring the exponent down to the front of the log.

$$\frac{1}{2}(\log 5 + \log x + \log y)$$ Use the product rule to expand the log.

b. Combining these logarithms will also use the product and quotient rules for logarithms.

i.

$$\log 8 + 3\log x + 4\log y$$

$$\log 8 + \log x^3 + \log y^4$$

Use the power rule to bring the coefficients into the logs as exponents.

$$\log 8x^3y^4$$

Combine the logs using the product rule.

ii.

$$\ln 3 + 2\ln 5a - \ln b$$

$$\ln 3 + \ln(5a)^2 - \ln b$$

$$\ln 3 + \ln(25a^2) - \ln b$$

Use the power rule to bring the coefficient into the log as an exponent. Both the 5 and a are now squared. Simplify the constants.

$$\ln\left(\frac{75a^2}{b}\right)$$

Combine the logs using the product and quotient rules.

iii. $\ln\left(\frac{xy^2}{z^5}\right)$

iv. $\log_7\left(\frac{10m}{n^3}\right)$

EXAMPLE ② PRACTICE PROBLEM

a. Expand the following logarithms as separate simpler logarithms without any exponents.

i. $\log_4(5xy^3)$

ii. $\log\left(\frac{3m^2}{2n^3}\right)$

iii. $\ln(\sqrt{2xy})$

b. Write the following logarithms as a single logarithm with a coefficient of one.

i. $2\log_5 m + 3\log_5 n + \log_5 7 - 2\log_5 p$

ii. $\log x + \frac{1}{2}\log y - 3\log z$

iii. $5\ln x + 3\ln 5z + 5\ln y - 5\ln 2$

These rules for logarithms together with the basic rules discussed in the Section 6.2 allow us to work with logarithms, expanding or simplifying them as needed

6.4 Exercises

For Exercises 1 through 12, expand the logarithms as separate simpler logarithms with no exponents.

1. $\log(5xy^2)$

2. $\ln(2h^2k^3)$

3. $\log_7(3x^3y^4)$

4. $\log_3\left(\frac{1}{2}ab^2c^3\right)$

5. $\log\left(\frac{2x^2}{y}\right)$

6. $\log\left(\frac{5x^3y}{2z^2}\right)$

7. $\ln\left(\frac{3x^4y^3}{z}\right)$

8. $\log_9(2w^2z^3)$

9. $\log_{15}\left(\frac{\sqrt{3x^4y^3}}{z^5}\right)$

10. $\log_9(\sqrt{2w^5z^3})$

11. $\ln(\sqrt[5]{m^2p^3})$

12. $\log_7(\sqrt[3]{3x^3y^4})$

For Exercises 13 through 30, write the logarithms as a single logarithm with a coefficient of one.

13. $\log 5 + 2\log x + \log y + 3\log z$

14. $\log 3 + 3\log x + 4\log y - \log 7$

15. $\log_3 a + 2\log_3 b - 5\log_3 c$

16. $\ln 5 + 2\ln x + 3\ln z - 4\ln y$

17. $\log_5 7 + \frac{1}{2}\log_5 x + \frac{1}{2}\log_5 y$

18. $\log 4 + \frac{1}{2}\log a - \frac{1}{2}\log b$

19. $\frac{1}{2}\ln 7 + \frac{1}{2}\ln a + \frac{3}{2}\ln b - 4\ln c$

20. $\log_9 5 + \frac{2}{3}\log_9 x + \frac{2}{3}\log_9 y$

21. $5\ln 7 + 2\ln a + 4\ln b - 3\ln c - 2\ln d$

22. $\log 3x + 3\log x - 2\log 7y$

23. $\log_3 a^2 + 2\log_3 bc - 5\log_3 3c$

24. $\ln 5x^3 + 2\ln x + 3\ln z - 4\ln yz$

25. $\log_5 7 + 2\log_5 xy + \log_5 xy$

26. $\frac{1}{2}\ln 7 + \frac{1}{2}\ln 5a + \frac{3}{2}\ln ab - 3\ln c$

27. $\log(x+6) + \log(x-2)$

28. $\log(x+1) + \log(x-5)$

29. $\ln(x-3) + \ln(x-7)$

30. $\ln(x-10) + \ln(x+4)$

Solving Exponential Equations

SOLVING EXPONENTIAL EQUATIONS

In this section we will use the rules for exponentials and logarithms to solve problems involving exponentials. There are two basic techniques used to solve exponential equations. You will find that the power rule for logarithms and the change of base formula are some of the most important rules that we use in the solving process. Recall the following rules for exponents and logarithms.

RULES FOR EXPONENTS

1. $x^m \cdot x^n = x^{m+n}$
        2. $(xy)^m = x^m y^m$

3. $\dfrac{x^m}{x^n} = x^{m-n}$
        4. $\left(\dfrac{x}{y}\right)^m = \dfrac{x^m}{y^m}$

5. $(x^m)^n = x^{mn}$
        6. $x^0 = 1$

7. $x^{-1} = \dfrac{1}{x}$
        8. $x^{1/n} = \sqrt[n]{x}$

RULES FOR LOGARITHMS

1. $\log_b(mn) = \log_b m + \log_b n$
    2. $\log_b(m^n) = n \log_b m$

3. $\log_b\left(\dfrac{m}{n}\right) = \log_b m - \log_b n$
    4. $\log_b x = \dfrac{\log x}{\log b}$

5. $\log_b(b) = 1$
    6. $\log_b(1) = 0$

Remember that when working with equations, what we do to one side of an equation we must also do to the other side of the equation. Because a logarithm is a one-to-one function, we can take the logarithm of both sides of an equation much as we would square both sides. When taking the logarithm of both sides, you are not multiplying, but each side is now the input to the logarithm. Taking the logarithm of both sides of an exponential equation and then using the power rule for logarithms is one option. Writing an exponential equation in logarithm form and then using the change of base formula is another option when solving. In all of these problems we use either log or ln, since these are logarithms that we can calculate on our calculator.

EXAMPLE   SOLVING AN EXPONENTIAL EQUATION
2 OPTIONS

Solve $4^x = 20$.

Skill Connection

When working with exponential equations, it is best to use a logarithm base that is convenient to calculate and find approximate values for.

It is good to note that within any problem you solve, you can use any base logarithm you wish, but you should stay consistent throughout the problem itself.

SC-Example 1: Solve

$$9^x = 62$$

Solution:

Using log:

$$\log(9^x) = \log 62$$
$$x \log 9 = \log 62$$
$$x = \frac{\log 62}{\log 9}$$
$$x \approx 1.878$$

Using ln:

$$\ln(9^x) = \ln 62$$
$$x \ln 9 = \ln 62$$
$$x = \frac{\ln 62}{\ln 9}$$
$$x \approx 1.878$$

In this example,

$$x = \frac{\log 62}{\log 9} = \frac{\ln 62}{\ln 9}$$

are exact answers, since they have not been rounded during calculation.

$$x \approx 1.878$$

is an approximate solution, since it has been rounded after calculation.

In some circumstances it is best to leave an exact answer. In most of our applications the exact answer is difficult to interpret, so we use the approximate solution.

Solution

Option 1 Take the log of both sides Use the power rule for logs		Option 2 Rewrite in log form Use the change of base formula	
$4^x = 20$ $\log(4^x) = \log(20)$ $x \log 4 = \log 20$ $\dfrac{x \log 4}{\log 4} = \dfrac{\log 20}{\log 4}$ $x = \dfrac{\log 20}{\log 4}$ $x \approx 2.161$	Take the log of both sides. Use the power rule for logs to bring down the variable x. Divide both sides by log 4.	$4^x = 20$ $x = \log_4 20$ $x = \dfrac{\log 20}{\log 4}$ $x \approx 2.161$	Rewrite in log form. Use the change of base formula.

Either of these methods work well for solving exponential equations. Using rewriting in logarithm form and using the change of base formula can lead to fewer mistakes with more complicated equations so we will use this option more often.

In section 6.2 we introduced the change of base formula but never demonstrated how it can be derived using the rules for logarithms. If we take $\log_3 400 = x$ we can use the same solving technique used in option 1 of Example 1 to show how the change of base formula comes about.

$$\log_3 400 = x$$ We first rewrite the log in exponential form.

$$3^x = 400$$

$$\log(3^x) = \log 400$$ Take the log of both sides of the equation and use the power rule for logs to bring the exponent down.

$$x \log 3 = \log 400$$

$$x = \frac{\log 400}{\log 3}$$ Divide both sides by log 3 to isolate x. This last result is the same as the change of base formula.

This same solving technique can be done without numbers to give us the change of base formula in abstract form as in the definition found in section 6.2.

STEPS TO SOLVING EXPONENTIAL EQUATIONS (OPTION 1)

1. Isolate the base and exponent on one side of the equation.
2. Take the logarithm of both sides of the equation.
3. Use the power rule for logarithms to bring the exponent down to the front of the log.
4. Isolate the variable.

> **STEPS TO SOLVING EXPONENTIAL EQUATIONS (OPTION 2)**
>
> 1. Isolate the base and exponent on one side of the equation.
> 2. Rewrite the exponential in logarithm form.
> 3. Use the change of base formula to write the log in a convenient base.
> 4. Isolate the variable.

EXAMPLE   2   SOLVING AN EXPONENTIAL MODEL

As the population ages, more and more people will develop the symptoms of dementia resulting from Alzheimer's disease. According to medical studies shown at www.Brain.com, the number of people worldwide who suffer from Alzheimer's disease could double in the next 25 years. Data from past numbers and projections for the future are given in the bar graph.

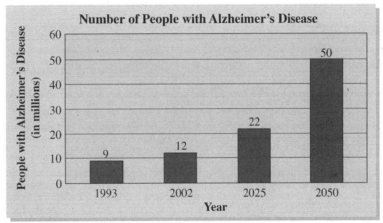

Source: Data estimated from information given at www.brain.com.

a. Find a model for these data.

b. Estimate the number of people with Alzheimer's disease in 2010.

c. Find when there will be 30 million people with Alzheimer's.

Solution

a. First we need to define the variables.

   A = The number of people in the world with Alzheimer's disease in millions

   t = Time in years since 1990

With these definitions we get the following adjusted data.

t	A(t)
3	9
12	12
35	22
60	50

These data give us the following scatterplot.

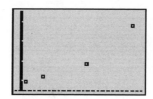 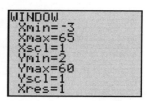

Because of the description given in the problem and the sudden rise in the data, we can choose an exponential model. Using the points (12, 12) and (60, 50), we get

$$(12, 12) \quad \text{and} \quad (60, 50)$$

$$12 = a \cdot b^{12} \qquad 50 = a \cdot b^{60}$$
Use the two points to write two equations in standard form.

$$\frac{50}{12} = \frac{a \cdot b^{60}}{a \cdot b^{12}}$$
Divide the two equations and solve for b.

$$4.167 = b^{48}$$

$$(4.167)^{1/48} = (b^{48})^{1/48}$$
Use the power rule for exponents to isolate b.

$$1.03 = b$$

$$12 = a(1.03)^{12}$$

$$12 = 1.426a$$

$$\frac{12}{1.426} = \frac{1.426a}{1.426}$$

$$8.415 = a$$

$$A(t) = 8.415(1.03)^t$$

This model gives us the graph.

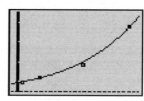

This model seems to fit reasonably, so we will use it.

b. 2010 would be represented by $t = 20$, so we get

$$A(20) = 15.198$$

Thus in 2010 there will be approximately 15 million people in the world with Alzheimer's disease.

c. 30 million people with Alzheimer's disease would be represented by $A(t) = 30$, so we get

$$30 = 8.415(1.03)^t$$

$$\frac{30}{8.415} = \frac{8.415(1.03)^t}{8.415}$$
Start by isolating the exponential part.

$$3.565 = 1.03^t$$

$$\log_{1.03} 3.565 = t \qquad \text{Rewrite the exponential in logarithm form.}$$

$$\frac{\log 3.565}{\log 1.03} = t \qquad \text{Use the change of base formula.}$$

$$43 \approx t$$

Check the answer using the calculator.

Therefore we find that in 2033 there will be approximately 30 million people worldwide with Alzheimer's disease. This of course assumes that the current trend continues and that no effective cures or treatments are discovered by then.

As you see from this example, when the variable you are trying to solve for is in the exponent, you will need to bring that variable down in order to isolate it on one side of the equation. This is why the power rule for logarithms is so crucial. By taking the logarithm of both sides of the equation, we can use the power rule for logarithms to bring the variable in the exponent down and make it multiplication. It is crucial to first get the exponential part of the equation by itself before you take the logarithm of both sides.

Using Your TI Graphing Calculator

When dividing a number by more than one quantity, you will need to use parentheses for the calculator to perform the operation you want done.

To calculate the fraction

$$\frac{\log 5.6}{3 \log 5}$$

you will need to put parentheses around the entire denominator.

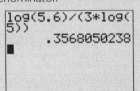

Without the parentheses around the denominator the calculator will follow the order of operations and first divide by 3 and then multiply by log 5.

EXAMPLE 3 SOLVING EXPONENTIAL EQUATIONS

Solve the following exponential equations.

a. $5^{3x} = 5.6$

b. $3^x - 9 = 21$

c. $2(8)^{x-2} = 24$

d. $10^{x^2 - 2} = 100$

Solution

a.
$$5^{3x} = 5.6 \qquad \text{The base and exponent are already isolated.}$$

$$\log_5 5.6 = 3x \qquad \text{Rewrite the exponential in logarithm form.}$$

$$\frac{\log 5.6}{\log 5} = 3x \qquad \text{Use the change of base formula.}$$

$$\frac{\log 5.6}{3 \log 5} = x \qquad \text{Isolate } x \text{ by dividing both sides by 3.}$$

$$x \approx 0.3568$$

Check the answer using the calculator.

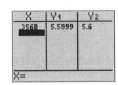

b. $3^x - 9 = 21$ Isolate the exponential part.

$$\underline{ +9 + 9 }$$

$$3^x = 30$$ Take the logarithm of both sides.

$$\ln(3^x) = \ln 30$$

$$x \ln 3 = \ln 30$$ Use the power rule for logarithms to bring down the exponent.

$$\frac{x \ln 3}{\ln 3} = \frac{\ln 30}{\ln 3}$$

                     Solve for x.

$$x \approx 3.0959$$

X	Y₁	Y₂
3.0959	21	21

X=

Check the answer using the calculator.

c. $2(8)^{x-2} = 24$

$$\frac{2(8)^{x-2}}{2} = \frac{24}{2}$$ Isolate the base and exponent by dividing both sides by 2.

$$8^{x-2} = 12$$

$$x - 2 = \log_8 12$$ Rewrite in logarithm form.

$$x - 2 = \frac{\log 12}{\log 8}$$ Use the change of base formula.

$$\underline{ +2 +2 }$$ Isolate x.

$$x = \frac{\log 12}{\log 8} + 2$$ This is the exact solution.

$$x \approx 3.195$$ This is an approximate solution.

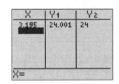

X	Y₁	Y₂
3.195	24.001	24

X=

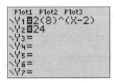

```
Plot1  Plot2  Plot3
\Y1■2(8)^(X-2)
\Y2■24
\Y3=
\Y4=
\Y5=
\Y6=
\Y7=
```

Check the answer using the calculator. Be sure to use parentheses around the exponent.

d. $10^{x^2-2} = 100$

$$x^2 - 2 = \log 100$$ Rewrite using logarithm form.

$$x^2 - 2 = 2$$ Evaluate the logarithm.

$$x^2 = 4$$ We have a quadratic left to solve.

$$x = \pm\sqrt{4}$$ Remember to use the plus/minus symbol when using the square root property.

$$x = \pm 2$$

X	Y₁	Y₂
2	100	100
-2	100	100

X=

```
Plot1  Plot2  Plot3
\Y1■10^(X2-2)
\Y2■100
\Y3=
\Y4=
\Y5=
\Y6=
\Y7=■
```

Check the answer using the calculator. Be sure to use parentheses around the exponent.

EXAMPLE ③ PRACTICE PROBLEM

Solve the following exponential equations.

a. $2^{5x} - 12 = 7$ **b.** $7^{2x+3} = 81$

EXAMPLE ④ SOLVING AN EXPONENTIAL APPLICATION

In Section 5.5 we were give the following model for the California sea otter population.

$$O(t) = 31.95(1.057)^t$$

where $O(t)$ represents the number of California sea otters t years since 1900.

Source: Model derived from data from the National Biological Service.

According to this model, when were there 1500 California sea otters?

Solution

1500 California sea otters are represented by $O(t) = 1500$, so we get

$$1500 = 31.95(1.057)^t$$

$$\frac{1500}{31.95} = \frac{31.95(1.057)^t}{31.95} \qquad \text{Isolate the base and exponent.}$$

$$46.95 = 1.057^t$$

$$\log 46.95 = \log 1.057^t \qquad \text{Take the logarithm of both sides.}$$

$$\log 46.95 = t \log 1.057 \qquad \text{Use the power rule for logarithms.}$$

$$\frac{\log 46.95}{\log 1.057} = t \qquad \text{Isolate the variable } t.$$

$$69.4 \approx t$$

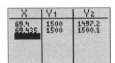

Check the answer using a calculator.
Use more decimal places to confirm that the solution is correct.

So we know that in 1969 there were approximately 1500 California sea otters.

EXAMPLE ④ PRACTICE PROBLEM

In Section 5.5 we found the following model for the population of Boulder, Colorado.

$$P(t) = 96000(1.1)^t$$

where $P(t)$ represents the population of Boulder, Colorado, t years since 2000.

Source: Model derived from data in Census 2000.

According to this model when will the population of Boulder, Colorado, reach 1 million?

COMPOUNDING INTEREST

Recall from Section 5.5 the following formulas for compounding interest problems

COMPOUNDING INTEREST

$$A = P\left(1 + \frac{r}{n}\right)^{nt}$$

A = The amount in the account

P = The principal (the amount initially deposited)

r = The annual interest rate written as a decimal

n = The number of times the interest is compounded in one year

t = The time the investment is made for, in years

COMPOUNDING CONTINUOUSLY

$$A = Pe^{rt}$$

In many situations we might want to know how long it will take to double an investment or at what rate we need to invest money in order for it to double in a certain amount of time. The compounding interest formulas can be used to answer these questions. When solving for time, we will be solving an exponential equation. When solving for the rate, we will be solving a power equation. When solving the equation for compounding continuously, we would be solving an exponential equation.

EXAMPLE **5** FIND A RATE GIVEN DOUBLING TIME

JP Manufacturing has $100,000 to invest. They need to double their money in 10 years to replace a piece of equipment that they rely on for their business. Find the annual interest rate compounded daily that they need in order to double their money in the 10 years.

Solution

In this case we have the following given amounts.

$100,000 is to be invested, so $P = 100000$.

The investment is for 10 years, so $t = 10$.

They need to double their money, so the account balance will need to be $200,000, so $A = 200000$.

The interest rate is to compound daily, so $n = 365$.

This leaves us to find the interest rate r, so we need to solve a power equation.

$$200000 = 100000\left(1 + \frac{r}{365}\right)^{365(10)} \qquad \text{Substitute the values of } P, n, A, \text{ and } t.$$

$$200000 = 100000\left(1 + \frac{r}{365}\right)^{3650} \qquad \text{Simplify the exponent.}$$

$$\frac{200000}{100000} = \frac{100000\left(1 + \dfrac{r}{365}\right)^{3650}}{100000}$$

Isolate the parentheses.

$$2 = \left(1 + \frac{r}{365}\right)^{3650}$$

$$2^{1/3650} = \left(\left(1 + \frac{r}{365}\right)^{3650}\right)^{1/3650}$$

Use a reciprocal exponent to eliminate the exponent.

$$1.000189921 = 1 + \frac{r}{365}$$

Solve for *r*.

$$1.000189921 = 1 + \frac{r}{365}$$

When you subtract 1 the calculator will usually give you the result in scientific notation.

$$\frac{-1 \qquad\qquad -1}{0.000189921 = \dfrac{r}{365}}$$

$$365(0.000189921) = 365\left(\frac{r}{365}\right)$$

$$0.0693 \approx r$$

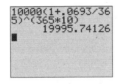

Check the answer.

   Therefore JP Manufacturing needs to find an account that will pay at least 6.93% compounded daily to double the initial amount in 10 years.

EXAMPLE 6 FIND THE DOUBLING TIME GIVEN AN INTEREST RATE

Many people use U.S. government savings bonds as a long-term investment. When you purchase a series I or EE savings bond, you pay half of the face value of the bond. The bond matures when it has earned enough interest to be worth the entire face value of the bond. If you invested $4000 in series EE bonds, how long would it take the bonds to mature if the bonds pay an average of 4% interest compounded semiannually?

Solution

Because we invested $4000, we will want the bonds to grow until the value doubles that amount, so we have the following values.

$$P = 4000 \qquad A = 8000$$
$$r = 0.04 \qquad n = 2$$

We need to find the time it takes to double the investment, so we need to solve for t.

$$8000 = 4000\left(1 + \frac{0.04}{2}\right)^{2t}$$

$$2 = \left(1 + \frac{0.04}{2}\right)^{2t} \qquad \text{Isolate the base and exponent.}$$

$$2 = 1.02^{2t} \qquad \text{Simplify the base.}$$

$$\log_{1.02} 2 = 2t \qquad \text{Rewrite in logarithm form.}$$

$$\frac{\log 2}{\log 1.02} = 2t \qquad \text{Use the change of base formula.}$$

$$\frac{\frac{\log 2}{\log 1.02}}{2} = \frac{2t}{2} \qquad \text{Isolate the variable.}$$

$$\frac{\frac{\log 2}{\log 1.02}}{2} = t$$

$$17.5 \approx t$$

 Check the answer.

So it will take about 17.5 years for the $4000 to double at 4% compounded semiannually.

EXAMPLE ⑥ PRACTICE PROBLEM

How long will it take for $10,000 to triple if it is invested in an account that pays 5% compounded weekly?

EXAMPLE 7 FIND THE DOUBLING TIME GIVEN CONTINUOUSLY COMPOUNDED INTEREST RATE

If $2000 is invested in a savings account that pays 5% annual interest compounded continuously, How long will it take for the investment to double?

Solution

We know the following values.

$$P = 2000 \qquad r = 0.05 \qquad A = 4000$$

t is the missing quantity that we are being asked to solve for.

$$4000 = 2000e^{0.05t}$$

$$2 = e^{0.05t}$$

$$\ln 2 = 0.05t$$

$$\frac{\ln 2}{0.05} = t$$

$$13.86 \approx t$$

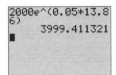

Check the answer.

Therefore it will take about 13.86 years for this investment to double.

6.5 Exercises

1. The use of supercomputers at universities all over the country has been a vital part of research and development for many areas of education. The Cray C90 housed at Rutgers University in New Jersey has seen dramatic increases in its use per year since 1987. The number of hours the Cray C90 has been used per academic year are given below.

Academic Year	Cray C90 Hours
1987	32
1988	100
1989	329
1990	831
1991	1,685
1992	2,233
1993	3,084
1994	8,517
1995	15,584
1996	27,399

Source: Rutgers University High Performance Computing.

a. Find a model for these data.

b. Give a reasonable domain and range for this model.

c. Estimate the number of hours the Cray C90 was used in 1997.

d. According to your model, when will the Cray C90 be used 500,000 hours per year?

2. The number of weeds in the author's backyard started to decline rapidly after being treated with weed killer. The approximate number of weeds remaining is recorded below.

Days	Weeds Remaining
1	900
3	345
5	130
6	80
7	50

a. Find a model for these data.

b. Give a reasonable domain and range for this model.

c. Estimate the number of weeds remaining after 4 days.

d. According to your model, when will there be only 10 weeds remaining?

3. According to United Nation's records, the world population in 1975 was approximately 4 billion. Statistics indicate that the world population since World War II has been growing at a rate of 1.9% per year. Assuming exponential growth, the world population can be modeled by

$$P(t) = 4(1.019)^t$$

where $P(t)$ is the world population in billions and t is the time in years since 1975. When will the world population reach 10 billion?

4. In Exercise 15 of Section 5.5 you found a model such as

$$W(t) = 350(1.14)^t$$

Where $W(t)$ is the number of humpback whales off the coast of Australia t years since 1981. According to this model, when will the whale population reach 10,000?

5. The white-tailed deer population in the northeastern United States has been growing since the early 1980s. The population of white-tailed deer can be modeled by

$$W(t) = 1.19(1.08)^t$$

where $W(t)$ represents the number of white-tailed deer in millions in the Northeast t years since 1980. When will the white-tailed deer population reach 5 million?

Source: Model derived from data from the National Biological Service.

6. Dr. Arnd Leike, a professor of physics at the University of Munich in Germany, won a 2002 Ig Nobel Prize in Physics for his investigation into the "exponential decay" of a beer's head (the foam on top when poured). After testing several beers for the rate of decay of the head, he came to the conclusion that a beer could be identified by its unique head decay rate. The following functions are for three of the beers that Dr. Leike tested.

$$E(s) = 16.89(0.996)^s$$
$$B(s) = 13.25(0.991)^s$$
$$A(s) = 13.36(0.993)^s$$

where $E(s)$ represents the height of the head of an Erdinger beer, $B(s)$ is the height of the head of a Budweiser Budvar, and $A(s)$ is the head height of an Augustinerbrau. All head heights are in centimeters s seconds after being poured.

Source: European Journal of Physics, Volume 23, 2002.

a. You might not want to drink a beer until the head has decayed to only 1 cm. How long will it take the head on the Erdinger beer to reach that height?

b. How long will it take the head on the Budweiser Budvar beer to decay to only 1 cm?

c. How long will it take the head on the Augustinerbrau beer to decay to only 1 cm?

7. In Section 5.5 Example 4 we found the following model for the population of South Africa.

$$P(t) = 43.8(0.995)^t$$

Where $P(t)$ is the population of South Africa in millions t years since 2008.

a. According to this model when will South Africa's population be only 40 million?

b. If this trend continues, how long will it take South Africa's population to be half of what it was in 2008?

8. In Section 5.5 Exercise 17 we found a model for the population of Denmark

$$P(t) = 5.5(1.00295)^t$$

where $P(t)$ is the population of Denmark in millions t years since 2008.

a. According to this model when will Denmark's population reach 6 million?

b. How long will it take Denmark's population to double if it continues to grow at this rate?

9. Cogs R Us has $40,000 to invest. They need to double their money in eight years to replace a piece of equipment that they rely on for their business. Find the annual interest rate compounded daily that they need in order to double their money in the eight years.

10. John and Mary hope to pay for their child's college education. They have $20,000 to invest and believe that they will need about 4 times that much money in about 17 years to pay for college. Find the annual interest rate compounded daily that they need in order to have the money they need in the 17 years.

11. What interest rate compounded continuously do you need to earn to make $5000 double in 12 years?

12. What interest rate compounded continuously do you need to earn to make $7000 double in 9 years?

13. What interest rate compounded monthly do you need to earn to make $50,000 double in 7 years?

14. What interest rate compounded monthly do you need to earn to make $46,000 double in 10 years?

15. How long will it take an investment of $5000 to double if it is deposited in an account paying 7% interest compounded monthly?

16. How long will it take an investment of $4000 to double if it is deposited in an account paying 5% interest compounded daily?

17. How long will it take an investment of $10,000 to double if it is deposited in an account paying 9% interest compounded continuously?

18. How long will it take an investment of $8,000 to double if it is deposited in an account paying 4.5% interest compounded continuously?

19. How long will it take an investment of $8000 to triple if it is deposited in an account paying 2% interest compounded daily?

20. How long will it take an investment of $4000 to triple if it is deposited in an account paying 6% interest compounded daily?

21. How long will it take an investment of $7000 to triple if it is deposited in an account paying 2.5% interest compounded continuously?

22. How long will it take an investment of $12,000 to triple if it is deposited in an account paying 5.75% interest compounded continuously?

For Exercises 23 through 40, solve the following exponential equations. Round all answers to three decimal digits.

23. $3^w = 125$

24. $7^x = 240$

25. $4^{x+7} = 3$

26. $5^{x+2} = 235$

27. $5(3.2)^x = 74.2$

28. $4(1.5)^x = 30.375$

29. $2^{2t} - 58 = 6$

30. $3.4^t + 8 = 47.304$

31. $-3(2.5)^m - 89 = -3262$

32. $-5(1.4)^x - 548 = -6475$

33. $3(2)^{n+2} = 96$

34. $5(4)^{3x+2} - 830 = 14.5$

35. $2^{x^2+6} = 1024$

36. $10^{x^2-5} = 0.10$

37. $3(4.6)^{2m+7} = 17.3$

38. $7(1.8)^{3x+5} = 12.8$

39. $5(4.3)^x + 7 = 2(4.3)^x + 89$

40. $7(2.5)^x + 12 = 4(2.5)^x + 49$

In Exercises 41 through 46, find the inverse for the given function.

41. $f(x) = 5(3)^x$

42. $g(x) = 3.4(10)^x$

43. $h(x) = -2.4(4.7)^x$

44. $h(x) = -3.5(1.8)^x$

45. $g(x) = -3.4e^x$

46. $f(x) = 4.2e^x$

For Exercises 47 through 52, solve the system of equations graphically.

47. $y = 2.5(3)^x$
$y = 150(0.8)^x$

48. $y = 2(5)^x$
$y = 300(0.6)^x$

49. $y = -5(0.8)^x$
$y = -1.8x - 20$

50. $y = 1.5(1.8)^x$
$y = 3x + 15$

51. $y = 105(0.7)^x$
$y = -1.5x^2 + 5x + 50$

52. $y = -24(1.2)^x$
$y = 3x^2 + 2x - 75$

Solving Logarithmic Equations

APPLICATIONS OF LOGARITHMS

In this section we will learn to use the rules for both exponentials and logarithms to solve equations involving logarithms. Logarithms are used most often in areas of science that require us to work with very large or extremely small numbers.

Solving a logarithm equation is very similar to solving an exponential equation except that you will rewrite the logarithm into exponential form instead of the other way around.

STEPS TO SOLVING LOGARITHMIC EQUATIONS

1. Isolate the logarithm(s) on one side of the equation.
2. Combine logarithms into a single logarithm if necessary.
3. Rewrite the logarithm in exponential form.
4. Isolate the variable.

One area that logarithms are used is the measurement of earthquakes. The size of an earthquake is most often measured by its magnitude. In 1935 Charles Richter defined the magnitude of an earthquake to be

$$M = \log\left(\frac{I}{S}\right)$$

where I is the intensity of the earthquake (measured by the amplitude in centimeters of a seismograph reading taken 100 km from the epicenter of the earthquake) and S is the intensity of a "standard earthquake" (whose amplitude is 1 micron $= 10^{-4}$ cm). (A standard earthquake is the smallest measurable earthquake known.)

© Image State/Alamy

EXAMPLE   **1**   EARTHQUAKE'S MAGNITUDE

If the intensity of an earthquake were 40,000 cm, what would its magnitude be?

Solution

The intensity of this earthquake is 40,000 cm, so $I = 40,000$. $S = 10^{-4}$ cm, so we can substitute these values into the magnitude formula and calculate the magnitude.

$$M = \log\left(\frac{40000}{10^{-4}}\right)$$

$$M \approx 8.6$$

Therefore an earthquake that has an intensity of 40,000 cm has a magnitude of 8.6.

EXAMPLE   **2**   EARTHQUAKE'S INTENSITY

What is the intensity of an earthquake of magnitude 6.5?

Solution

Because we know the intensity of a standard earthquake is 10^{-4}, we can substitute this for S and 6.5 for M and solve for the intensity of the earthquake.

$$6.5 = \log\left(\frac{I}{10^{-4}}\right)$$

$$10^{6.5} = \frac{I}{10^{-4}} \qquad \textbf{Rewrite the logarithm in exponential form.}$$

$$10^{-4}(10^{6.5}) = \left(\frac{I}{10^{-4}}\right)10^{-4} \quad \textbf{Add the exponents together.}$$

$$10^{2.5} = I$$

$$316.23 \approx I$$

Check the answer using the calculator table.

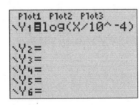

Therefore the intensity of a magnitude 6.5 earthquake is about 316.23 cm. A magnitude 6.5 earthquake hit central California in 2003, killing two people and destroying several buildings. If this earthquake had hit in a more developed area, the destruction could have been much worse.

EXAMPLE   **2**   PRACTICE PROBLEM

What is the intensity of an earthquake with magnitude 8?

Another application is the use of common logarithms in chemistry to calculate the pH of a solution. The pH of a solution is a measurement of how acidic or alkaline a solution is. A neutral solution will have a pH value of 7, and an acidic solution like

Skill Connection

In chemistry hydrogen ion concentrations are typically written using scientific notation. In Example 4 we find the hydrogen ion concentration

$$10^{-4.7} = H^+$$

To convert this to scientific notation we will calculate this in the calculator.

The E in the result indicates scientific notation and is written as

$$1.995 \times 10^{-5}$$

The number after the E becomes the exponent of 10.

vinegar (pH of about 3) or your stomach acid (pH of about 1) will have pH values less than 7. Alkaline solutions such as lye (pH of about 9), used to make soap, have pH values greater than 7. The pH of a solution can be calculated by taking the negative log of the hydrogen ion concentration.

$$pH = -\log(H^+)$$

The hydrogen ion concentration has the unit M which stands for the molarity of the solution.

EXAMPLE 3 FIND PH OF ACIDS AND BASES

What is the pH of an aqueous solution when the concentration of hydrogen ion is 5.0×10^{-4} M?

Solution

Because the hydrogen ion concentration is given as 5.0×10^{-4} M, we can substitute that in for H^+ and get

$$pH = -\log(5.0 \times 10^{-4})$$

$$pH = 3.30$$

Therefore an aqueous solution with a hydrogen ion concentration of 5.0×10^{-4} M has a pH of 3.30.

EXAMPLE 4 FIND HYDROGEN ION CONCENTRATION

Find the hydrogen ion concentration of the solution in the beaker shown.

Solution

Because we know the pH value, we can substitute 4.7 for pH in the formula and solve for H.

$$4.7 = -\log(H^+)$$ **Multiply both sides by -1 to isolate the logarithm.**

$$-4.7 = \log(H^+)$$

$$10^{-4.7} = H^+$$ **Rewrite the logarithm in exponential form.**

$$1.995 \times 10^{-5} = H^+$$ **Write the solution using scientific notation.**
    Check the solution using the calculator table.

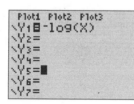

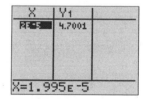

Therefore a solution with a pH value of 4.7 will have a hydrogen ion concentration of 1.995×10^{-5} M.

EXAMPLE PRACTICE PROBLEM

Find the hydrogen ion concentration of a solution with pH = 2.3.

SOLVING OTHER LOGARITHMIC EQUATIONS

EXAMPLE 5 SOLVE LOGARITHMIC EQUATIONS

Solve the following logarithm equations.

a. $\log x = \log 5$

b. $\log(x + 20) = 2$

c. $\ln t = 3$

d. $\log(2x) + 7 = 12$

Solution

a. In this case we have a logarithm that is equal to another logarithm of the same base, so we know that the inside of the logarithms must be equal. Therefore $x = 5$ is the solution to this equation.

b. This equation starts out with a logarithm isolated on the left side of the equation, and it is equal to a constant. In this case we can rewrite the logarithm in exponential form and then solve.

$$\log(x + 20) = 2$$
$$10^2 = x + 20 \qquad \text{Rewrite the logarithm into exponential form.}$$
$$100 = x + 20$$
$$80 = x$$
$$\log(80 + 20) = 2 \qquad \text{Check the solution.}$$
$$\log(100) = 2$$

c. Again we have a single logarithm equal to a constant, so we rewrite the logarithm into exponential form and solve.

$$\ln t = 3$$
$$e^3 = t \qquad \text{Rewrite the logarithm into exponential form.}$$
$$20.086 \approx t$$

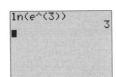

Check the solution.

In this case e^3 is the exact answer and is sometimes the required or needed solution. 20.086 is an approximate answer and is suitable for many applications.

d. We must first start by isolating the logarithm on one side of the equation, and then we rewrite the logarithm into exponential form and solve.

$$\log(2x) + 7 = 12$$

$$\underline{\;-7\;-7} \qquad \text{Isolate the logarithm}$$

$$\log(2x) = 5$$

$$10^5 = 2x \qquad \text{Rewrite the logarithm into exponential form.}$$

$$100000 = 2x \qquad \text{Solve for } x.$$

$$50000 = x$$

Check the solution.

X	Y₁	Y₂
50000	12	12

X=

EXAMPLE ⑤ PRACTICE PROBLEM

Solve the following logarithm equations.

a. $\log x = 3.5$

b. $\log(x + 4) = 4$

c. $\ln(5x) + 8 = 3$

You can see that the relationship between logarithms and exponentials will play a big role in how we solve these equations. In general, when trying to solve an equation that has a variable inside a logarithm, you will want to isolate the logarithm and then rewrite the logarithm in exponential form. This will get the variable out of the logarithm and allow you to solve the equation. If there are more than one logarithm in the equation, first combine them into a single log using the rules for logarithms.

EXAMPLE ⑥ SOLVE LOGARITHM EQUATIONS

Solve the following logarithm equations.

a. $\log_6(x) + \log_6(2x) = 2$

b. $\log(2x) + \log(x - 4) = 1$

c. $\log_2(3x^2) - \log_2 x = 5$

d. $\log_2(x + 5) + \log_2(x + 4) = 1$

Solution

a. First you can combine the logarithms together using the product rule for logarithms and then rewrite the logarithm in exponential form so you can solve for x.

$$\log_6(x) + \log_6(2x) = 2$$

$$\log_6(2x^2) = 2 \qquad \text{Combine the logarithms.}$$

$$6^2 = 2x^2 \qquad \text{Rewrite the logarithm into exponential form.}$$

$$36 = 2x^2$$

$$18 = x^2$$

$$\pm\sqrt{18} = x \qquad \text{These two answers will need to be checked.}$$

$$\pm 4.24 = x$$

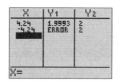 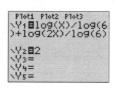

Check the solutions. Use the change of base formula to enter the logarithms.

In this case we have two possible solutions, but because we cannot evaluate the logarithm of a negative number, $x = -4.24$ cannot be a solution. Therefore $x = 4.24$ is our only valid solution.

b. Again combine the logarithms and then rewrite in exponential form.

$$\log(2x) + \log(x - 4) = 1$$

$$\log(2x^2 - 8x) = 1 \qquad \text{Combine the logarithms.}$$

$$10 = 2x^2 - 8x \qquad \text{Rewrite the logarithm into exponential form.}$$

$$0 = 2x^2 - 8x - 10$$

$$0 = 2(x - 5)(x + 1) \qquad \text{We have a quadratic left, so we can factor or use the quadratic formula.}$$

$$x - 5 = 0 \qquad x + 1 = 0$$

$$x = 5 \qquad x = -1 \qquad \text{These two answers will need to be checked.}$$

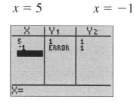

Again we have two possible solutions, but when $x = -1$ is substituted into the logarithm, it is impossible to evaluate. Therefore $x = 5$ is our only solution.

c. In this situation we will need to use the quotient rule for logarithms to combine the two logarithms that are subtracted.

$$\log_2(3x^2) - \log_2 x = 5$$

$$\log_2\left(\frac{3x^2}{x}\right) = 5 \qquad \text{Simplify the fraction.}$$

$$\log_2(3x) = 5$$

$$2^5 = 3x \qquad \text{Rewrite into exponential form.}$$

$$32 = 3x$$
$$10.67 \approx x$$

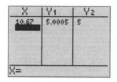 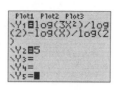   Check the solution.

d. $\log_2(x + 5) + \log_2(x + 4) = 1$

$\log_2(x^2 + 9x + 20) = 1$ **Combine the logarithms using the product**

$x^2 + 9x + 20 = 2^1$ **rule and then rewrite in exponential form.**

$x^2 + 9x + 18 = 0$ **Factor.**

$(x + 3)(x + 6) = 0$

$x + 3 = 0 \qquad x + 6 = 0$

$x = -3 \qquad x = -6$

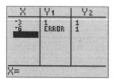 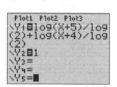   Check the solutions.

Although we found two answers for this problem, only $x = -3$ works in this case. When you substitute $x = -6$ into the equation, it is undefined. The negative 3 does work, because once it is substituted, you get a positive value inside each logarithm.

EXAMPLE ⑥ PRACTICE PROBLEM

Solve the following logarithm equations.

a. $\log(2x) + \log(3x) = 3$ **b.** $\ln(3x + 5) = 4$

c. $\log_3(x + 11) + \log_3(x + 5) = 3$ **d.** $\log_4(3x) + \log_4(2x - 7) = 2$

Exercises

1. If the intensity of an earthquake were 2000 cm, what would its magnitude be?

2. In May 2003 Japan had an earthquake that had an intensity of 1000 cm. What was the magnitude of this earthquake?

3. If the intensity of an earthquake were 500,000 cm, what would its magnitude be?

4. The strongest recorded earthquake was in Chile in 1960 and had an intensity of 316227.77 cm. What was its magnitude?

5. The deadliest earthquake ever recorded had an estimated 830,000 fatalities. This occurred in January 1556 in Shensi, China. What was the intensity of this earthquake if its magnitude was estimated as 8.0?

6. Many of the strongest earthquakes in the world take place off the coast of Alaska in the Andreanof Islands or in Prince William Sound. In 1957 a 9.1 magnitude earthquake hit this area. What was the intensity of that earthquake?

7. The deadliest earthquake of 2003 was in Southeastern Iran and had a magnitude of 6.6. What was its intensity?

8. On May 12, 2008 China had a 7.9 Magnitude earthquake. What was the intensity of that earthquake?

9. What is the pH of Scope mouthwash if its hydrogen ion concentration is 1.0×10^{-7} M?

10. What is the pH of the aqueous solution shown in the beaker?

Hydrogen Ion Concentration 4×10^{-7}

11. Find the hydrogen ion concentration of car battery acid which has a pH = 1.

12. Find the hydrogen ion concentration of Pepsi, which has a pH = 3.

13. Find the hydrogen ion concentration of an oven cleaner that has a pH = 14.

14. Find the hydrogen ion concentration of a healthy person's stomach acid, which has a pH around 1.36.

15. Blood plasma should have a pH level between 7.35 and 7.45. Find the hydrogen ion concentrations that would fall in this pH range.

16. Swiss cheese 1 day into the production should have a pH level between 5.2 and 5.4. Find the hydrogen ion concentrations that would ball in this pH range.

For Exercises 17 through 40, solve each logarithm equation. Check that all solutions are valid.

17. $\log(5x + 2) = 2$

18. $\log(2x + 7) = 3$

19. $\ln(3t) + 5 = 3$

20. $\ln(2t) + 8 = 2$

21. $\log_5(3x) + \log_5 x = 4$

22. $\log_6(4x) + \log_6 x = 2$

23. $\log_3(x + 2) + \log_3(x - 3) = 2$

24. $\log_2(2x) + \log_2(5x - 4) = 6$

25. $\log(5x^3) - \log(2x) = 2$

26. $\log_4(6x^3) - \log_4(3x) = 5$

27. $\ln(x + 4) = 2$

28. $\log_4(x - 2) = 2$

29. $\log(2x + 1) = -0.5$

30. $\log(x - 3) = -2$

31. $\log_5(x) + \log_5(2x + 7) - 3 = 0$

32. $\log_5(x + 12) + \log_5(x + 8) - 2 = -1$

33. $\log_4(-2x + 5) + \log_4(x + 21.5) = 4$

34. $\log_2(-x + 4) + \log_2(x + 3.5) = 3$

35. $3 \log_2(2x) + \log_2(16) = 7$

36. $\ln(3x) + \ln(x + 5) = 4$

37. $\ln(3x^2 + 5x) - 3 = 2$

38. $\log(x^2 - 3x) - 4 = 1$

39. $2 \log(3x) + 8 = 2$

40. $5 \log(2x) + 4 = 3$

Chapter 6 Summary

Section 6.1 Introduction to Inverses

- An **inverse function** will "undo" what the original function did to the input values.

- To find an inverse function within a context, solve the function for the original input variable and rewrite using function notation showing the two variables changing from input to output and vice versa.

- To find an inverse function when not in a context, follow these steps.

    1. Take it out of function notation. (Replace $f(x)$ with y in the equation.)
    2. Solve for the original input variable.
    3. Interchange the variables. This will keep x as the input variable.
    4. Rewrite in function notation using the inverse notation $f^{-1}(x)$.

- **One-to-one functions** have exactly one input for each output. All one-to-one functions have inverses.

- You can test whether a function is one-to-one by using the horizontal line test.

- The domain and range of inverse functions will be the same, but switched. The domain becomes the range of the inverse, and the range becomes the domain.

EXAMPLE

The number of minutes an hour a local radio station can play music depends on the number of commercials played during that hour. Math Rocks 101.7 FM uses the following equation when considering how much time they have to play music.

$$M(C) = 60 - 0.5C$$

where M is the number of minutes of music the station can play in an hour if it plays C 30-second commercials that hour.

a. Find an inverse for this function.

b. Find the number of commercials the radio station can play if it wants to have 50 minutes of music each hour.

c. If the domain for the original function is [2, 30], find the domain and range for the inverse function.

Solution

a.
$$M(C) = 60 - 0.5C$$
$$M = 60 - 0.5C$$
$$M - 60 = 0.5C$$
$$-2M + 120 = C$$
$$C(M) = -2M + 120$$

b. $C(50) = 20$, so if the radio station wants to play 50 minutes of music each hour, it can play 20 thirty-second commercials each hour.

c. Domain: [45, 59]. Range: [2, 30].

EXAMPLE 2

Find the inverse for the function $g(x) = -3x + 5$

Solution

$$g(x) = -3x + 5$$
$$y = -3x + 5$$
$$y - 5 = -3x$$
$$-\frac{y}{3} + \frac{5}{3} = x$$
$$-\frac{x}{3} + \frac{5}{3} = y$$
$$g^{-1}(x) = -\frac{x}{3} + \frac{5}{3}$$

Section 6.2 Introduction to Logarithms

- A **logarithm** is defined as the inverse of an exponential function. The basic form of a logarithmic function is $f(x) = \log_b x$, where b is called the **base** of the logarithm and b must be positive and not equal to 1.

- A logarithm asks the question "What exponent of the base b will give me what's inside?"

- Some basic rules for logarithms are as follows.

  $\log_b 1 = 0$. The logarithm of 1 is always zero.

  $\log_b b = 1$. The logarithm of its own base is always 1.

  $\log_b(b^m) = m$. The logarithm of its base to a power is just that power.

- The **common logarithm** is log base 10. This logarithm is written as "log" with no base noted.

- The **natural logarithm** is log base e. This logarithm is written as "ln" with no base noted.

- The **change of base formula** can be used to convert a logarithm from one base to another: $\log_b x = \dfrac{\log_c x}{\log_c b}$.

- Note that c is typically 10 or e, since these logs are on your calculator.

- Logarithms and exponentials are inverses of one another and can be used to solve equations. Often, rewriting a logarithm into exponential form or an exponential into logarithm form will help solve an equation.

EXAMPLE   3

Evaluate the following logarithms.

a. $\log_3 27$

b. $\log 275$

c. $\ln 13$

d. $\log_6 145$

Solution

a. $\log_3 27 = 3$

b. $\log 275 = 2.439$

c. $\ln 13 = 2.565$

d. $\log_6 145 = \dfrac{\log 145}{\log 6} = 2.778$

EXAMPLE   4

Solve the logarithmic equation by rewriting it into exponential form.

a. $\log(3x) = 2$

b. $\ln(x + 5) = 4$

c. $\log_4(x) = 3$

Solution

a. $\log(3x) = 2$

$$10^2 = 3x$$

$$\frac{100}{3} = x$$

b. $\ln(x + 5) = 4$

$$e^4 = x + 5$$

$$e^4 - 5 = x$$

$$49.6 = x$$

c. $\log_4(x) = 3$

$$4^3 = x$$

$$64 = x$$

Section 6.3 Graphing Logarithmic Functions

- The basic characteristics of logarithmic graphs are:

  No y-intercept.

  x-intercept of $(1,0)$.

  The y-axis is a vertical asymptote.

When the base is greater than 1, the graph increases quickly for inputs close to zero and then more slowly as the inputs increase.

When the base is between 0 and 1, the graph decreases rapidly for inputs close to zero and then more slowly as the inputs increase.

- The domain of basic logarithmic functions is all positive real numbers or $(0, \infty)$.

- The range of basic logarithmic functions is all real numbers, or $(-\infty, \infty)$.

EXAMPLE 5

Sketch the graph of $f(x) = \log_5 x$.

Solution

Because this logarithm has a base of 5, the inverse will be the exponential function $f^{-1}(x) = 5^x$. Using this exponential function to create a table of points we will plot the graph.

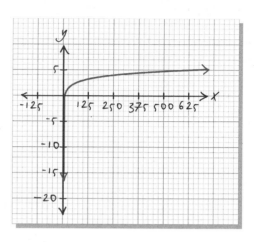

Points for a graph of the inverse.

x	$f^{-1}(x) = 5^x$
-1	0.2
0	1
1	5
2	25
3	125
4	625

Points for the logarithm graph.

x	$f(x) = \log_5 x$
0.2	-1
1	0
5	1
25	2
125	3
625	4

Section 6.4 Properties of Logarithms

- The three main properties of logarithms are as follows.

  The product rule for logarithms:
  $\log_b(mn) = \log_b m + \log_b n.$

  The quotient rule for logarithms:
  $\log_b\left(\dfrac{m}{n}\right) = \log_b m - \log_b n.$

  The power rule for logarithms:
  $\log_b(m^n) = n \log_b m.$

- Using these properties for logarithms allows us either to break a complicated logarithm apart into smaller logarithms or to combine several logarithms of the same base together as a single log.

EXAMPLE 6

Expand the following logarithms into separate simpler logarithms with no exponents.

a. $\log(5x^2y)$

b. $\ln\left(\dfrac{4x^3y^2}{z^5}\right)$

c. $\log_5(\sqrt{6x^3yz})$

Solution

a. $\qquad \log(5x^2y)$

$\qquad \log 5 + \log x^2 + \log y$

$\log 5 + 2\log x + \log y$

b. $\ln\left(\dfrac{4x^3y^2}{z^5}\right)$

$\ln 4 + \ln x^3 + \ln y^2 - \ln z^5$

$\ln 4 + 3 \ln x + 2 \ln y - 5 \ln z$

c. $\log_5(\sqrt{6x^3yz})$

$\dfrac{1}{2} \log_5 (6x^3yz)$

$\dfrac{1}{2}(\log_5 6 + \log_5 x^3 + \log_5 y + \log_5 z)$

$\dfrac{1}{2}(\log_5 6 + 3 \log_5 x + \log_5 y + \log_5 z)$

EXAMPLE 7

Combine the following logarithms into one logarithm with a coefficient of 1.

a. $\log 7 + 4 \log x + 2 \log y - \log z$

b. $\ln 3 + \dfrac{1}{2}\ln x + \ln y$

Solution

a. $\log 7 + 4 \log x + 2 \log y - \log z = \log\left(\dfrac{7x^4y^2}{z}\right)$

b. $\ln 3 + \dfrac{1}{2}\ln x + \ln y = \ln(3y\sqrt{x})$

Section 6.5 Solving Exponential Equations

- When **solving exponential equations,** use one of the following solving techniques.

  Get the exponential term by itself, take the logarithm of both sides of the equation, and then use the power rule for logarithms to make the exponent a coefficient. This will allow you to finish solving the equation.

  Get the exponential term by itself, and rewrite into logarithm form. Use the change of base formula if needed and finish solving for the variable.

EXAMPLE 8

Solve the following equations.

a. $5 \cdot 3^{4x} = 75$

b. $4^{2x+1} + 7 = 115$

Solution

a. $5 \cdot 3^{4x} = 75$

$3^{4x} = 15$

$\log(3^{4x}) = \log 15$

$4x \log 3 = \log 15$

$x = \dfrac{\log 15}{4\log 3}$

$x \approx 0.616$

Check the answer.

X	Y₁	Y₂
.616	74.92	75

X=

b. $4^{2x+1} + 7 = 115$

$4^{2x+1} = 108$

$\log(4^{2x+1}) = \log 108$

$(2x + 1)\log(4) = \log 108$

$2x + 1 = \dfrac{\log 108}{\log 4}$

$2x = \dfrac{\log 108}{\log 4} - 1$

$x = \dfrac{\dfrac{\log 108}{\log 4} - 1}{2}$

$x \approx 1.1887$

Check the answer.

X	Y₁	Y₂
1.1887	114.99	115

X=

Section 6.6 Solving Logarithm Equations

- When solving logarithmic equations follow this basic approach.

  Combine all logarithms into a single logarithm, using the log rules.

  Isolate the logarithm.

  Rewrite the logarithm into exponential form.

  Finish solving the equation.

  Check your solution in the original problem.

- The magnitude of an earthquake can be found using the formula $M = \log\left(\dfrac{I}{S}\right)$, where M is the magnitude of the earthquake on the Richter scale, I is the intensity of the earthquake in centimeters, and S is the intensity of a standard earthquake, 10^{-4} cm.

- The pH of a solution can be found using the formula $\text{pH} = -\log(H^+)$, where H^+ is the hydrogen ion concentration of the solution.

EXAMPLE 9

Solve the following equations.

a. $\ln(3x + 5) = 4$

b. $\log(5x) + \log(2x + 7) = 2$

Solution

a. $\ln(3x + 5) = 4$

$$e^4 = 3x + 5$$

$$\frac{e^4 - 5}{3} = x$$

$$16.53 = x$$

Check the answer.

X	Y₁	Y₂
16.53	3.9999	4

X=

b. $\log(5x) + \log(2x + 7) = 2$

$$\log(10x^2 + 35x) = 2$$
$$10x^2 + 35x = 100$$
$$10x^2 + 35x - 100 = 0$$
$$x = -5.36 \qquad x = 1.86$$
$$x = 1.86$$

Check the answer.

X	Y₁	Y₂
-5.36	ERROR	2
1.86	1.9987	2

X=

Only one answer works this time. The -5.36 makes the inside of a log negative and thus it cannot be evaluated.

EXAMPLE 10

The pH of vinegar is 3. Find the hydrogen ion concentration.

Solution

$$3 = -\log(H^+)$$
$$-3 = \log(H^+)$$
$$10^{-3} = H^+$$

The hydrogen ion concentration of vinegar is 0.001 M.

Chapter 6 Review Exercises

For Exercises 1 through 5, find the inverse for each function.

1. $f(x) = 1.4x - 7$ **[6.1]**

2. $g(t) = -2t + 6$ **[6.1]**

3. $h(x) = 5^x$ **[6.2]**

4. $f(x) = 3.5(6)^x$ **[6.2]**

5. $g(x) = 2e^x$ **[6.2]**

6. The AEM Toy Company can manufacture
$D(e) = 400e + 50$ dolls per day when e employees
are working.

    a. Find an inverse for the given function.

    b. How many dolls can the AEM Toy Company
manufacture if 10 employees are working on a
particular day?

    c. How many employees must work for the com-
pany to manufacture 1000 dolls in a day?

    d. If the company can employe between 5 and 30
employees in a day, give a domain and range for
the inverse function.   **[6.1]**

For Exercises 7 and 8, sketch a graph of the function.

7. $f(x) = \log_{2.5}(x)$ **[6.3]**

8. $f(x) = \log_{0.4}(x)$ **[6.3]**

9. Give the domain and range of the function in Exercise
7.   **[6.3]**

10. Give the domain and range of the function in Exercise
8.   **[6.3]**

*For Exercises 11 through 13, write the logarithms as
separate simpler logarithms with no exponents.*

11. $\ln(2x^3y^4)$ **[6.4]**

12. $\log_3(\sqrt{5ab^2c^3})$ **[6.4]**

13. $\log\left(\dfrac{3x^3y^5}{z^4}\right)$ **[6.4]**

*For Exercises 14 through 17, write the logarithms as a
single logarithm with a coefficient of one.*

14. $\log 4 + 3 \log x + 2 \log y + \log z$ **[6.4]**

15. $\log 2 + 4 \log x + 5 \log y - 2 \log z$ **[6.4]**

16. $\log_3 a + 3 \log_3 b - 2 \log_3 c$ **[6.4]**

17. $\ln 7 + 2 \ln x + \ln z - 5 \ln y - 3 \ln z$ **[6.4]**

18. The number of Franklin's gulls in North America has
been in significant decline in the past several years. The
population of Franklin's gulls can be modeled by

$$F(t) = 750(0.9405^t)$$

where $F(t)$ represents the number of Franklin's
gulls in thousands in North America t years
since 2000.

Source: Model derived from data from the National Biological Service.

    a. Estimate the Franklin's gull population in 2005.

    b. When will the Franklin's gull population reach
500,000?

    c. What might cause this population to stop
declining at this rate?   **[6.5]**

19. How long will it take an investment of $7000 to
double if it is deposited in an account that pays 6%
interest compounded monthly?   **[6.5]**

20. How long will it take an investment of $100,000 to dou-
ble if it is deposited in an account that pays 8.5% interest compounded continuously?   **[6.5]**

*For Exercises 21 through 25, solve the following exponen-
tial equations.*

21. $4^w = 12.5$ **[6.5]**

22. $7^{x+2} = 47$ **[6.5]**

23. $17(1.9)^x = 55.4$ **[6.5]**

24. $3(5)^{3x+2} - 700 = 11$ **[6.5]**

25. $5^{x^2+4} = 15625(6)$ **[6.5]**

26. What was the intensity of the magnitude 5.7 earthquake that hit Round Valley, California, in November 1984?   **[6.6]**

27. What is the pH of cranberry juice if its concentration of hydrogen ion is 3.7×10^{-3} M?   **[6.6]**

28. Find the hydrogen ion concentration of a shampoo with a pH = 5.6.   **[6.6]**

For Exercises 29 through 33, solve each logarithm equation. Check that all solutions are valid.

29. $\log(10x + 20) = 5$ **[6.6]**

30. $\ln(4.3t) + 7.5 = 3$ **[6.6]**

31. $\log_5(2x) + \log_5 x = 4$ **[6.6]**

32. $\log_3(x - 4) + \log_3(x + 1) = 3$ **[6.6]**

33. $\log(8x^5) - \log(2x) = 3$ **[6.6]**

Chapter 6 Test

1. How long will it take an investment of $1000 to double if it is deposited in an account that pays 3.5% interest compounded monthly?

2. What was the intensity of the magnitude 7.0 earthquake that hit Papua, Indonesia, in February 2004?

For Exercises 3 through 6, solve the equation.

3. $\log_{16}(2x + 1) = -0.5$

4. $\log_7 x + \log_7 (2x + 7) - 3 = 0$

5. $20 = -3 + 4(2^x)$

6. $8 = 3^{7x-1}$

7. Is this the graph of a one-to-one function? Explain why or why not.

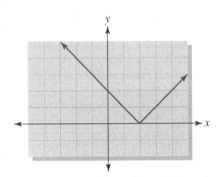

8. Write as a single logarithm with coefficient 1.

$$4 \log(2a^3) + 5 \log (ab^3)$$

9. Write the logarithm as separate simpler logarithms with no exponents.

$$\log\left(\frac{3xy^5}{\sqrt{z}}\right)$$

10. Sketch the graph of $f(x) = \log_{1.25} x$

11. The number of CD singles shipped in the mid-1990s is given in the following table.

Year	CD Singles (in millions)
1993	7.8
1994	9.3
1995	21.5
1996	43.2
1997	66.7

a. Find a model for these data.

b. Give a reasonable domain and range for this model.

c. Estimate the number of CD singles shipped in 2000.

d. When were only 5 million CD singles shipped?

Source: Statistical Abstract 2001.

For Exercises 12 through 15, find the inverse of the function.

12. $f(x) = 5x + 2$

13. $g(x) = -2.4x - 7.3$

14. $f(x) = 5^x$

15. $h(x) = 2(7)^x$

16. Find the hydrogen ion concentration of a solution with pH = 2.58.

Chapter 6 Project

HOW FAST IS THAT TUMOR GROWING?

Written Project

One or more people

In this project you will be asked to investigate the growth of an average breast cancer tumor. A simple model of the growth of a cancerous tumor is one in which the number of cancer cells doubles on average every 100 days. Current mammography technology can detect a tumor that is 1 cubic centimeter (cc) in size. A tumor that is 1 cc consists of about 1 billion cells. A tumor that is 1 liter (1000 cc) is often considered lethal. Use this information to answer the following questions.

Write up

a. If a tumor starts from a single cell that has undergone a malignant transformation, find a model for the number of cells that are present as time goes by.

b. Create a graph of your model on the calculator or computer and print it out or draw it very neatly by hand on graph paper. Remember that we are going to look at very large numbers of cells, so scale your graph carefully.

c. Use your model to estimate how long after that initial malignant transformation occurred until the tumor is detectable by using a mammogram.

d. If the tumor is not detected or treated, how long will it take for the tumor to reach a typically lethal size?

e. What is the range in time that mammography can detect a tumor before it reaches a lethal size?

f. With this time range in mind, how often should a woman get a mammogram? Explain your reasoning.

Rational Functions

- Identify a rational function.
- Find a rational model from simple contexts.
- Understand and find the domain and range of a rational function in a context.
- Interpret direct and indirect variation.
- Simplify rational expressions.
- Use long division of polynomials.
- Use synthetic division.
- Multiply and divide rational expressions.
- Add and subtract rational expressions.
- Simplify complex fractions.
- Solve rational equations.

© Ingram Publishing/Alamy

t is extremely important to the survival of ships near the world's coastlines that lighthouses give off intense light that can be seen through the night and through inclement weather. The light from a lighthouse must be seen from far away to be effective, but the intensity of light lessens the greater the distance a ship is from shore. In this chapter we will investigate inversely proportional relationships such as this one. One of the chapter projects will ask you to model the mechanics of a lighthouse by testing the intensity of light coming from a flashlight at various distances.

  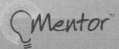

SECTION
7.1

Introduction to Rational Functions

RATIONAL FUNCTIONS

In this chapter we will study another type of function that is found in many areas of business, physics, and other sciences. Rational functions are functions that contain fractions involving polynomials. These functions can be simple or very complex. Rational functions often arise from combining two functions together using division. In this section we will learn some of the basics about different characteristics of rational functions and areas in which they occur.

CONCEPT INVESTIGATION I ARE WE THERE YET? ··

Let's start by considering a simple situation involving driving a car. John is planning to take a 200-mile car trip. Let's consider how long this trip will take John if he travels at different average speeds.

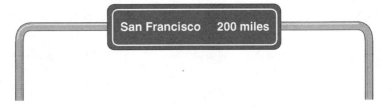

San Francisco   200 miles

a. Fill in the following table with the times it will take John to travel 200 miles if he drives at an average rate of r miles per hour.

Rate (r) (in mph)	Time (t) (in hours)
10	20
25	8
40	
50	
80	
100	
200	

What happens to the time it takes to travel the 200 miles as the average speed gets slower?

What happens to the time it takes to travel the 200 miles as the average speed gets faster?

b. On your graphing calculator, create a scattergram of the data in the table. Does the scattergram agree with your answers from part a?

c.   Solve the equation $D = rt$ for time and use that to write a model for the time it takes John to travel 200 miles. (Verify your model by graphing it with your data.)

d.   What are a reasonable domain and range for the model in this situation?

A function such as $t(r) = \dfrac{200}{r}$ is a simple rational function because it has a variable in the denominator of the fraction. Any expression of the form

$$\frac{P(x)}{Q(x)}$$

where $P(x)$ and $Q(x)$ are polynomials and $Q(x) \neq 0$ is called a **rational expression.** Notice that $Q(x)$ cannot equal zero, or you would have division by zero, and thus the rational expression would be undefined. Division by zero is always a concern in working with rational expressions.

In Concept Investigation 1 you should note that it does not make sense for John to average zero miles per hour, or he will never travel the 200 miles.

DEFINITION

Rational expressions: An expression of the form

$$\frac{P(x)}{Q(x)}$$

where $P(x)$ and $Q(x)$ are polynomials and $Q(x) \neq 0$ is called a rational expression.

$$\frac{5x + 2}{x - 6} \quad \text{and} \quad \frac{3}{x}$$

are rational expressions.

In a context the domain of a rational function must avoid any values that make the denominator zero or turn the situation into something that does not make sense. The domain will usually be restricted to a small area of the graph, and thus the range will be the lowest to the highest points on the graph within the domain. Aa you work with a situation, set the domain according to what makes sense in the context. Be cautious with any restrictions stated in the problem that would limit the domain in some way.

EXAMPLE   1   PER PERSON COST

A group of students in the chess club wants to rent a bus to take them to the national chess competition. The bus is going to cost $1500 to rent and can hold up to 60 people.

a.   Find a model for the per person cost to rent the bus if s students take the bus and each student pays an equal share.

b.   How much would the per person cost be if 30 students take the bus?

c.   How much would the per person cost be if 60 students take the bus?

d.   What would a reasonable domain and range be for this model?

Solution

a. Let $C(s)$ be the cost per student in dollars for s students to take the bus to the national chess competition. Because each student is going to pay an equal amount, we might consider a few simple examples: If only one student takes the bus, that student would have to pay $1500. If two students take the bus, they will have to pay $\dfrac{1500}{2} = 750$ dollars each. So we are taking the total cost of $1500 and dividing it by the number of students taking the bus. This pattern would continue, and we would get the following function.

$$C(s) = \frac{1500}{s}$$

b. If 30 students take the bus, we can substitute 30 for s and calculate C.

$$C(30) = \frac{1500}{30}$$

$$C(30) = 50$$

Therefore if 30 students take the bus, it will cost $50 per person.

c. Substituting in $s = 60$, we get

$$C(60) = \frac{1500}{60}$$

$$C(60) = 25$$

Therefore if 60 students take the bus, it will cost $25 per person.

d. In an application problem we will continue to avoid model breakdown when setting a domain. Because the bus can hold only up to 60 people, we must limit the domain to positive numbers up to 60. This means that we could have a possible domain of [1, 60]. With this domain the range would be [25, 1500]. Of course, there are other possible domains and ranges, but these would be considered reasonable.

DIRECT AND INVERSE VARIATION

The function we found in example 1 part a is an example of **inverse variation,** and it could be stated that C varies inversely with s. That is, when one value increases, the other decreases. In example 1, the more students who take the bus, the lower the per student cost will be.

Variation occurs when two or more variables are related to one another using multiplication or division. When two variables are related and both either increase together or decrease together, we call it **direct variation.** The equation $D = 60t$ is an example of direct variation; when the value of t increases, so does the value of D.

Say What?

When it comes to variation there are a couple of phrases we use to describe the relationship between the two variables.

Direct Variation:

* varies directly with

* is directly proportional to

Inverse Variation:

* varies inversely with

* is inversely proportional to

DEFINITION

Direct variation: The variable y varies directly with x if

$$y = kx$$

Inverse variation: The variable y varies inversely with x if

$$y = \frac{k}{x}$$

In both cases k is called the variation constant (or the constant of proportionality). *Varies directly* and *varies inversely* can also be stated as *directly proportional to* and *inversely proportional to*, respectively.

The form of an equation that represents direct variation is a simple linear equation with a vertical intercept of (0, 0). The variation constant is the slope of the line. Because we studied such linear equations earlier, we will concentrate on problems involving inverse variation. You should note that an equation that represents variables that vary inversely is also a simple kind of rational function.

Say What?

The **illumination** (illuminance) of a light is the amount of light cast onto an object.

A **foot candle** is a unit of measure that originated with how much light a standard candle would cast on an object one foot away.

EXAMPLE | 2 | ILLUMINATION FROM A LIGHT SOURCE

The illumination of a light source is inversely proportional to the square of the distance from the light source. A certain light has an illumination of 50 foot-candles at a distance of 5 feet from the light source.

a. Find a model for the illumination of this light.

b. What is the illumination of this light at a distance of 10 feet from the light source?

c. What is the illumination of this light at a distance of 100 feet?

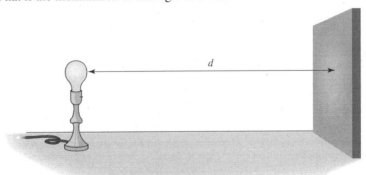

Solution

a. Let I be the illuminance of the light in foot-candles and let d be the distance from the light source in feet.

Because we are told that the illumination is inversely proportional to the square of the distance from the light source, the illumination will be equal to a constant divided by distance squared.

$$I = \frac{k}{d^2}$$

We still need to find the variation constant k. Because we are told that at 5 feet from the light source the illumination is 60 foot-candles, we can substitute these values in and find k.

$$50 = \frac{k}{5^2} \qquad \text{Substitute the given values for } I \text{ and } d.$$

$$50 = \frac{k}{25} \qquad \text{Multiply both sides by 25 to solve for } k.$$

$$1250 = k$$

Now that we know $k = 1250$, we can write the model for illumination as

$$I(d) = \frac{1250}{d^2}$$

b. We are given the distance of 10 feet, so we can substitute $d = 10$ and calculate the illumination.

$$I(10) = \frac{1250}{10^2}$$

$$I(10) = \frac{1250}{100}$$

$$I(10) = 12.5$$

Therefore the illumination at 10 feet from the light source is 12.5 foot-candles.

c. We are given the distance of 100 feet, so we can substitute $d = 100$ and calculate the illumination.

$$I(100) = \frac{1250}{100^2}$$

$$I(100) = \frac{1250}{10000}$$

$$I(100) = 0.125$$

Therefore the illumination at 100 feet from the light source is 0.125 foot-candle.

In any direct or inverse variation problem you need one example of the relationship between the two variables to find the variation constant. Then you will be able to write the formula for that particular situation. In Example 2 the variation constant will change on the basis of the strength of the light source you are working with.

EXAMPLE ② PRACTICE PROBLEM

In physics if a fixed amount force is applied to an object, the amount of acceleration a in meters per second squared of that object is inversely proportional to the mass in kilograms of that object. When a fixed amount of force is applied to a 5-kg mass, it is accelerated at 10 m/s^2 (m/s^2 is meters per second squared).

a. Find a function for the acceleration a of a mass m when this fixed amount of force is applied.

b. What is the acceleration of a 15-kg mass when this force is applied?

EXAMPLE 3 INVERSE VARIATION

a. If y varies inversely with $\sqrt{x}$ and $y = 58$ when $x = 4$, find an equation to represent this relationship.

b. Find y if $x = 9$.

Solution

a. Because y varies inversely with $\sqrt{x}$ we know that y will equal a constant divided by $\sqrt{x}$.

$$y = \frac{k}{\sqrt{x}}$$

We also know that $y = 58$ when $x = 4$ so we can substitute these values into y and x and solve for the constant k.

$$58 = \frac{k}{\sqrt{4}}$$

$$58 = \frac{k}{2}$$

$$116 = k$$

Now that we know $k = 116$, we can write the final equation

$$y = \frac{116}{\sqrt{x}}$$

b.   We will substitute $x = 9$ and find y.

$$y = \frac{116}{\sqrt{9}}$$

$$y = \frac{116}{3}$$

$$y = 38\frac{2}{3}$$

EXAMPLE ③ PRACTICE PROBLEM

a.   If y varies inversely with x^4 and $y = 143.75$ when $x = 2$, find an equation to represent this relationship.

b.   Find y if $x = 5$.

Another type of situation that involves rational functions occurs when we are combining two functions using division. In the next example we will investigate one of these situations.

EXAMPLE 4 FINDING AN AVERAGE

In the state of California the student/teacher ratio has been dropping steadily since about 1995. The total number of students in the K–12 public schools in California can be modeled by

$$S(t) = 112.25t + 4930.6$$

where S is the number of students in thousands in the California K–12 public schools and t is time in years since 1990. The number of teachers in California's K–12 public schools can be modeled by

$$T(t) = 11.75t + 175.6$$

where T is the number of teachers in thousands in the California K–12 public schools and t is time in years since 1990.

a.   Estimate the number of students in California's K–12 public schools in 1995.

b.   Estimate the number of teachers in California's K–12 public schools in 1995.

c.   Find a model for the average number of students per teacher in California's K–12 public schools.

d. Estimate the average number of students per teacher in California's K–12 public schools during the year 2000.

e. Use the graph of your new model to estimate when the number of students per teacher in California's K–12 public schools was 20.

Solution

a. We are given the year 1995, so we can substitute $t = 5$ and get

$$S(5) = 112.25(5) + 4930.6$$

$$S(5) = 5491.9$$

Therefore in 1995 there were approximately 5491.9 thousand students in California's K–12 public schools.

b. Again we are given 1995, but this time we substitute $t = 5$ into the $T(t)$ function.

$$T(5) = 11.75(5) + 175.6$$

$$T(5) = 234.35$$

Therefore in 1995 there were approximately 234.35 thousand teachers in California's K–12 public schools.

c. We are asked for a model that would give the average number of students per teacher in California's K–12 public schools, so we take the number of students and divide it by the number of teachers. Let $A(t)$ be the average number of students per teacher in California's K–12 public schools t years since 1990.

$$A(t) = \frac{S(t)}{T(t)} = \frac{112.25t + 4930.6}{11.75t + 175.6}$$

d. We are asked for the average number of students in 2000, so we can substitute $t = 10$ into our new model $A(t)$.

$$A(10) = \frac{112.25(10) + 4930.6}{11.75(10) + 175.6}$$

$$A(10) \approx 20.65$$

Therefore in 2000 there was an average of 20.65 students per teacher in California's K–12 public schools.

e. To find the graph of this function we can put the function into Y1 and then use the table to get an idea of the inputs that get us around the output of 20 that we are interested in.

Looking at the graph and using TRACE, we get the following.

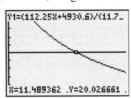

We can estimate that when $t \approx 11.5$, the number of students per teacher is about 20. Thus according to this model, around the year 2002 the average number of students per teacher in California's K–12 public schools was 20.

When working with problems like those in Example 3, you need to remember the rules for when you can divide two functions. When the functions are in context, be sure that the units of the input variables are the same and that when the units of the outputs are divided, the resulting unit makes sense in the context of the problem.

EXAMPLE PRACTICE PROBLEM

The number of people in the United States who are enrolled in Medicare can be modeled by

$$M(t) = 0.39t + 35.65$$

where $M(t)$ represents the number of people enrolled in Medicare in millions t years since 1990. Disbursements by the Medicare program can be modeled by

$$D(t) = -3160.64t^2 + 52536.5t - 4548.31$$

where $D(t)$ represents the amount of money Medicare disbursed to people in millions of dollars t years since 1990.

a. Find a model for the average amount disbursed per person enrolled in Medicare t years since 1990.

b. Estimate the average amount disbursed to a Medicare recipient in 1999.

c. Use the graph of your model to estimate the year when the average amount disbursed to Medicare recipients reached a maximum.

DOMAIN

When considering the domain of a rational function, we will mainly be concerned with excluding values from the domain that would result in the denominator's being zero. The easiest way to determine the domain of a rational function is to set the denominator equal to zero and solve. The domain then becomes all real numbers except those values that make the denominator equal zero.

Any place where the denominator is zero would result in a vertical asymptote, or a hole with a missing value. Recall that basic logarithmic functions had the y-axis as a vertical asymptote. Vertical asymptotes are also similar to the horizontal asymptotes that we saw in the graphs of exponential functions. The graph of a function will not touch a vertical asymptote but instead will get as close as possible and then jump over it and continue on the other side. Whenever an input value makes the numerator and denominator both equal to zero, a hole in the graph will occur, instead of a vertical asymptote. Determining when a hole or vertical asymptote occurs will be left for another course, but we will see them occur in the graphs of rational functions.

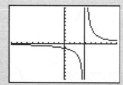
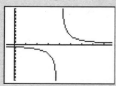
EXAMPLE 5 FINDING THE DOMAIN OF A RATIONAL FUNCTION

Find the domain of the following rational functions.

a. $f(x) = \dfrac{5}{x}$

b. $g(x) = \dfrac{5 + x}{x + 9}$

c. $h(x) = \dfrac{x + 4}{(x + 4)(x - 7)}$

d. $f(x) = \dfrac{3x + 2}{x^2 + 5x + 6}$

e.

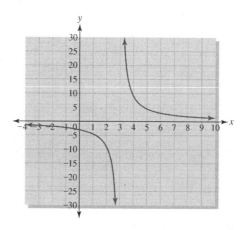

Solution

a. Because the denominator would be zero when $x = 0$, we have a domain of all real numbers except zero. This can also be written simply as $x \neq 0$. Looking at the graph of $f(x)$, we can see that the function jumps over the input value $x = 0$, and there is a vertical asymptote in its place.

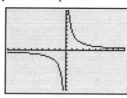

b. The denominator would be zero when $x = -9$, so its domain is all real numbers such that $x \neq -9$. Looking at this graph again, we see a vertical asymptote. Pay attention to the way in which this function must be entered into the calculator with parentheses around the numerator and another set of parentheses around the denominator of the fraction.

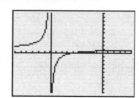

c. If you set the denominator equal to zero, you get

$$(x + 4)(x - 7) = 0$$

$$x + 4 = 0 \qquad x - 7 = 0$$

$$x = -4 \qquad x = 7$$

Therefore the domain is all real numbers except $x \neq -4$ or 7. This graph is shown in two parts so that you can see the hole that appears at $x = -4$ and then the asymptote at $x = 7$. Without setting up two windows, it is almost impossible to see the hole.

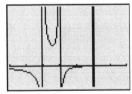 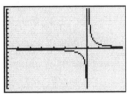

d. Set the denominator equal to zero.

$$x^2 + 5x + 6 = 0$$

$$(x + 3)(x + 2) = 0$$

$$x + 3 = 0 \qquad x + 2 = 0$$

$$x = -3 \qquad x = -2$$

Therefore the domain is all real numbers except $x \neq -3$ or -2. This graph has an interesting shape but does have two vertical asymptotes. Again the numerator and denominator of the fraction need parentheses around them to create the graph correctly.

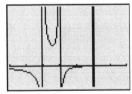 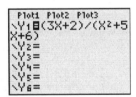

e. The graph of this function shows a vertical asymptote at about $x = 3$, so the domain should be all real numbers except $x \neq 3$.

EXAMPLE ⑤ PRACTICE PROBLEM

Find the domain of the following rational functions.

a. $f(x) = \dfrac{3}{x}$

b. $g(x) = \dfrac{x + 2}{x - 7}$

c. $h(x) = \dfrac{x + 2}{x^2 - 3x - 10}$

7.1 Exercises

1. Charity Poker Inc. organizes poker tournaments to raise money for charitable organizations. A typical tournament for up to 100 players costs $600 to put together.

   a. Find a model for the per player cost for a charity poker tournament if p players participate.

   b. Find the per person cost for each player if 75 players participate.

   c. What are a reasonable domain and range for this model?

2. The Adventure Guides group at a local YMCA is planning a camping trip on the beach. It costs $1500 to rent the campsites for the weekend and pay for insurance for up to 90 campers.

   a. Find a model for the per camper cost for this camp out if p people attend.

   b. Find the per person cost for each camper if 60 people attend.

   c. What are a reasonable domain and range for this model?

3. The Fancy Affair catering company is catering an event for a local charity. Jan, the owner of Fancy Affair, is donating her time to the cause but needs to charge the charity for the food, decorations, and supplies used. If p people attend the charity event, Jan has figured her total cost to be modeled by

$$C(p) = 4.55p + 365.00$$

where C is the total cost in dollars for the food, decorations, and supplies when p people attend the charity event.

   a. Find the total cost for 100 people to attend the event.

   b. Find the cost per person for 100 people to attend the event.

   c. Find a new function that gives the per person cost for p people to attend the charity event.

   d. Use your new model to find the per person cost for 150 people to attend the event.

   e. If the location of the charity event can only hold up to 250 people, what are a reasonable domain and range of the model you found in part b?

4. The math club is throwing a graduation party for five of its members. The club has decided to hire a band for $500 and to buy $300 worth of food and drinks for the party. Each member of the club who attends the party (except the five graduates) is to pay an equal amount of the costs.

   a. Find a model for the per person cost for this party if m members of the club attend the party. (m does include the five graduates.)

   b. Find the per person cost for each non graduating member of the club who attends the party if 45 members attend.

   c. What are a reasonable domain and range for this model if the math club has 50 members?

5. A light has an illumination of 25.5 foot-candles at a distance of 10 feet from the light source. (See Example 2.)

   a. Find a model for the illumination of this light.

   b. What is the illumination of this light at a distance of 5 feet from the light source?

   c. What is the illuminating of this light at a distance of 30 feet from the light source?

   d. Use a graph to estimate the distance at which the illumination will be 50 foot-candles.

6. If a fixed force is applied to an object, the amount of acceleration *a* in meters per second squared of that object is inversely proportional to the mass in kilograms of that object. When a fixed force is applied to a 20-kg mass, it is accelerated at 0.5 mps².

   a. Find a function for the acceleration *a* of a mass *m* when this fixed amount of force is applied.

   b. What is the acceleration of a 10-kg mass when this force is applied?

   c. What is the acceleration of a 5-kg mass when this force is applied?

   d. Estimate numerically what size mass this force can accelerate at 0.7 mps².

7. The weight of a body varies inversely as the square of its distance from the center of the earth. If the radius of the earth is 4000 miles, how much would a 220-lb man weigh 2000 miles above the surface of the earth? (*Hint:* The person weighs 220 lb when 4000 miles from the center of the earth.)

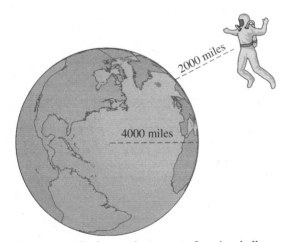

8. The pressure *P* of a certain amount of gas in a balloon is inversely proportional to the volume of the balloon. If the pressure in a balloon is 5 lb per square inch when the volume of the balloon is 4 cubic inches, what is the pressure in the balloon if the volume is only 2 cubic inches?

9. The current that flows through an electrical circuit is inversely proportional to the resistance of that circuit.

When the resistance *R* is 200 ohms, the current *I* is 1.2 amp. Find the current when the resistance is 130 ohms.

10. The force needed to balance a 200-g weight on a fulcrum is inversely proportional to the distance from the fulcrum at which the force is applied. 2.25 newtons of force applied 4 units away from the fulcrum is needed to balance the weight.

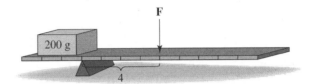

   a. Find the force needed to balance the 200-g weight if it is applied 1 unit away from the fulcrum.

   b. Find the force needed to balance the 200-g weight if it is applied 3 units away from the fulcrum.

   c. Find the force needed to balance the 200-g weight if it is applied 10 units away from the fulcrum.

11. The pressure of a certain amount of gas in a balloon is inversely proportional to the volume of the balloon. For one such balloon we have the function

$$P(v) = \frac{30}{v}$$

where $P(v)$ is the pressure of the gas in the balloon in pounds per square inch when the volume of the balloon is *v* cubic inches.

   a. Find $P(2)$ and explain its meaning in this context.

   b. Find $P(10)$ and explain its meaning in this context.

12. The force needed to lift a boulder using a pry-bar and fulcrum can be modeled by the function

$$F(d) = \frac{1500}{d}$$

where $F(d)$ is the force in newtons needed to lift the boulder when it is applied d inches away from the fulcrum.

a. Find $F(6)$ and explain its meaning in this context.

b. Find $F(120)$ and explain its meaning in this context.

13. a. If y varies inversely with x^3 and $y = 405$ when $x = 3$ find an equation to represent this relationship.

b. Find y if $x = 5$.

14. a. If y varies inversely with x^7 and $y = 1,867,776$ when $x = 4$ find an equation to represent this relationship.

b. Find y if $x = 6$.

15. a. If M is inversely proportional to $5\sqrt{t}$ and $M = 115$ when $t = 9$ find an equation to represent this relationship.

b. Find M if $t = 25$.

16. a. If y is inversely proportional to $8\sqrt{x}$ and $y = 25$ when $x = 6$ find an equation to represent this relationship.

b. Find y if $x = 4$.

17. The average amount of benefits received by people in the U.S. food stamp program can be modeled by

$$B(t) = \frac{-470001t^2 + 4110992t + 14032612}{-469.4t^2 + 3745t + 19774}$$

where $B(t)$ is the average benefit in dollars per person for people in the U.S. food stamp program t years since 1990.

Source: Model derived from data from Statistical Abstract 2001.

a. Find the average benefit for a person participating in the food stamp program in 1995.

b. Use a graph to estimate when the average benefit for a person participating in the food stamp program was $800.

18. The average amount of benefits received by people in the U.S. food stamp program can be modeled by

$$B(t) = \frac{-470001t^2 + 4110992t + 14032612}{-469.4t^2 + 3745t + 19774}$$

where $B(t)$ is the average benefit in dollars per person for people in the U.S. food stamp program t years since 1990.

Source: Model derived from data from Statistical Abstract 2001.

a. Estimate the average benefit in 2000.

b. Use a graph to estimate when the average benefit for a person participating in the food stamp program was $700.

19. The population of the United States since 1970 can be modeled by

$$P(t) = 0.226t^2 + 1.885t + 204.72$$

where $P(t)$ represents the population of the United States in millions t years since 1970. The national debt of the United States in millions of dollars t years since 1970 can be modeled by

$$D(t) = 8215.1t^2 - 23035.4t + 413525.6$$

where $D(t)$ represents the national debt of the United States in millions of dollars t years since 1970.

Source: Models derived from data from the U.S. Department of Commerce Bureau of Economic Analysis.

a. Find a new model for the average amount of national debt per person in the United States t years since 1970.

b. Find the average amount of national debt per person in 2000.

c. Estimate numerically the year in which the average amount of national debt per person was $10,000.

20. The state of California's spending per person has increased dramatically since 1950. The state's population from 1950 to 2000 can be modeled by

$$P(t) = 0.464t - 12.47$$

where $P(t)$ is California's population in millions of people t years since 1900. The amount that California has spent in millions of dollars can be modeled by

$$S(t) = 55.125t^2 - 6435.607t + 186914.286$$

where $S(t)$ is the amount California spent in millions of dollars t years since 1900.

Source: Models derived from data in the Governor's Budget Summary as printed in the North County Times, *Feb. 9, 2003.*

a.   Estimate the population of California in 1980.

b.   Estimate the amount California spent in 1990.

c.   Find a new function that gives the spending per capita (per person) t years since 1900.

d.   Estimate the per capita spending in California in 1980.

e.   Estimate numerically when the per capita spending in California reached $2500.

For Exercises 21 through 28, give the domain of the rational functions.

21. $f(x) = \dfrac{x + 5}{x - 3}$

22. $p(t) = \dfrac{3t - 7}{4t + 5}$

23. $h(a) = \dfrac{3a - 1}{(2a + 7)(a - 3)}$

24. $g(x) = \dfrac{2x - 7}{(x + 4)(x - 9)}$

25. $h(m) = \dfrac{4m^2 + 2m - 9}{m^2 + 7m + 12}$

26. $f(x) = \dfrac{3x + 7}{x^2 - 9x + 8}$

27. $f(x) = \dfrac{2x + 1}{x^2 + 3x + 19}$

28. $f(x) = \dfrac{x + 8}{x^2 + 5x + 21}$

For Exercises 29 and 32, use the graph to answer the questions.

29.   Given the graph of $f(x)$,

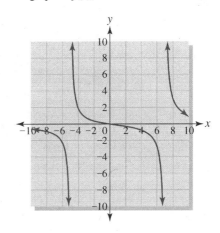

a.   Find the domain of $f(x)$.

b.   Estimate $f(0)$.

c.   Estimate x such that $f(x) = 5$.

30.   Given the graph of $f(x)$,

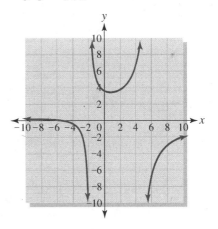

a.   Find the domain of $f(x)$.

b.   Estimate $f(0)$.

c.   Estimate x such that $f(x) = 6$.

31. Given the graph of $h(x)$,

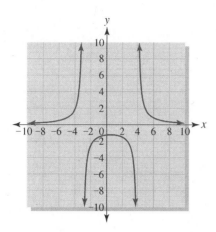

    a. Find the domain of $h(x)$.

    b. Estimate $h(1)$.

    c. Estimate $h(-4)$.

    d. Estimate x such that $h(x) = -2$.

32. Given the graph of $h(x)$,

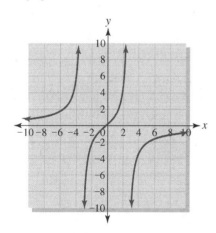

    a. Find the domain of $h(x)$.

    b. Estimate $h(8)$.

    c. Estimate $h(-6)$.

    d. Estimate x such that $h(x) = -8$.

For Exercises 33 and 38, use the table to give values that are not included in the function's domain.

33.

X	Y₁
-7	1.875
-6	5
-5	ERROR
-4	-15
-3	ERROR
-2	5
-1	1.875

X=-1

34.

X	Y₁
1	.66667
2	1.2
3	3
4	ERROR
5	-6
6	-6
7	ERROR

X=7

35.

X	Y₁
-6	-.1667
-4	ERROR
-2	0
0	-.0833
2	-.1667
4	-.375
6	ERROR

X=2

36.

X	Y₁
0	.33333
1	.875
2	3
3	ERROR
4	-13
5	ERROR
6	5.6667

37.

X	Y₁
-1	0
-2	-.3333
-3	ERROR
-4	3
-5	ERROR
-6	-1.667
-7	.75

X=-7

38.

X	Y₁
-40	.001
-30	.0025
-20	ERROR
-10	-.005
0	-.005
10	ERROR
20	.0025

X=20

Simplifying Rational Expressions

To prepare us to solve equations involving rational functions, we need to develop some basic skills to handle rational expressions. We are going to start by learning to simplify rational expressions using factoring and then look at long division of polynomials.

SIMPLIFYING RATIONAL EXPRESSIONS

Simplifying rational expressions is the same as simplifying numeric fractions. The numerator and denominator must both include multiplication in order for a factor to cancel and thus reduce the fraction. When working with numeric fractions, we often take this for granted and do the process without thinking about all the steps. To work with rational expressions, you will need to focus on each step of the process to properly reduce the expression.

Say What?

We often use "cancel out" in place of more specific phrases such as:

- Adds to zero
- Divides to 1
- Is the inverse of

This is usually done to simplify the language that is being used to make it easier to understand or read. Be cautious in using this phrase that you realize what operation is being used to "cancel" something out so that you do not "cancel out" at the wrong time.

SIMPLIFYING RATIONAL EXPRESSIONS

1. Factor the numerator and denominator (if needed).

2. Cancel any common factors.

3. Leave in factored form or multiply remaining factors together.

Caution: Do not cancel unless the expressions to be canceled are being multiplied by the remaining parts of the numerator or denominator. You can cancel factors, but you cannot cancel terms.

$$\frac{5\cancel{(x + 2)}}{(x + 3)\cancel{(x + 2)}} \qquad \frac{5 + (x + 2)}{(x + 3)(x + 2)}$$

Cancel.   Do not cancel.

Let's start with some simple examples with numeric fractions and some simple rational expressions.

EXAMPLE   I   SIMPLIFYING RATIONAL EXPRESSIONS

Simplify the following.

a. $\dfrac{25}{40}$ **b.** $\dfrac{420}{770}$

c. $\dfrac{5x^2}{10x}$ **d.** $\dfrac{2(x + 7)}{x + 7}$

e. $\dfrac{(x + 3)(x - 5)}{(x - 2)(x + 3)}$

Skill Connection

The most common place to cancel when you should not, is in division. We cannot cancel when there is addition involved rather than multiplication.

$$\frac{x+5}{x+2} \neq \frac{5}{2}$$

Here the x's cannot cancel using division because of the addition that is involved in the numerator and denominator.

$$\frac{5x}{2x} = \frac{5}{2}$$

In this fraction the x's can cancel out because only multiplication is involved, making the x's common factors that do divide to 1.

Solution

a. $\dfrac{25}{40} = \dfrac{5 \cdot 5}{2 \cdot 2 \cdot 2 \cdot 5}$ Factor the numerator and denominator.

$= \dfrac{5 \cdot \cancel{5}}{2 \cdot 2 \cdot 2 \cdot \cancel{5}}$ Cancel any common factors.

$= \dfrac{5}{2 \cdot 2 \cdot 2}$

$= \dfrac{5}{8}$ Multiply the remaining factors together.

b. $\dfrac{420}{770} = \dfrac{2 \cdot 2 \cdot 3 \cdot 5 \cdot 7}{2 \cdot 5 \cdot 7 \cdot 11}$ Factor the numerator and denominator.

$= \dfrac{2 \cdot 2 \cdot 3 \cdot \cancel{5} \cdot \cancel{7}}{\cancel{2} \cdot \cancel{5} \cdot \cancel{7} \cdot 11}$ Cancel any common factors.

$= \dfrac{2 \cdot 3}{11}$

$= \dfrac{6}{11}$ Multiply the remaining factors.

c. $\dfrac{5x^2}{10x} = \dfrac{5 \cdot x \cdot x}{2 \cdot 5 \cdot x}$ Factor the numerator and denominator.

$= \dfrac{\cancel{5} \cdot \cancel{x} \cdot x}{2 \cdot \cancel{5} \cdot \cancel{x}}$ Cancel any common factors.

$= \dfrac{x}{2}$

d. $\dfrac{2(x+7)}{x+7} = \dfrac{2\cancel{(x+7)}}{\cancel{x+7}}$ Already factored, so cancel out common factors.

$= \dfrac{2}{1}$ When all the factors in the numerator or denominator cancel, a 1 is left.

$= 2$ Reduce.

e. $\dfrac{(x+3)(x-5)}{(x-2)(x+3)} = \dfrac{\cancel{(x+3)}(x-5)}{(x-2)\cancel{(x+3)}}$ Cancel common factors.

$= \dfrac{x-5}{x-2}$ Note that the remaining x does not cancel because it is not multiplied but subtracted.

You should note that an expression involving addition or subtraction can be canceled but only if that expression is being multiplied by the remaining factors. Any addition or subtraction that remains separate from the multiplication in the numerator or denominator will stop you from canceling the like factors. The next example will work with some more complicated rational expressions, and we will consider a couple of situations in which we are not able to reduce the expression.

Skill Connection

The process for simplifying many rational expressions will require you to factor the numerator and denominator. Let's review the steps to factoring we went over in Section 3.4.

AC Method of Factoring:

1. Take out anything in common.
2. Multiply a and c together. (Do this step off to the side.)
3. Find factors of ac that add up to b. (Do this off to the side also.)
4. Rewrite the middle (bx) term using the factors from step 3.
5. Group and factor out what is in common.

SC-Example 1:

Factor $3x^2 - 13x - 10$.

Solution:

Step 1: Nothing in common.

Step 2:

$a = 3$ $c = -10$ $ac = -30$

Step 3:

$ac = -30$	-1	30
1	-30	-2	15
2	-15	-3	10
3	-10	-5	6
5	-6		

2 -15 gives us -13.

Step 4:

$3x^2 - 15x + 2x - 10$

Step 5:

$3x^2 - 15x + 2x - 10$

$(3x^2 - 15x) + (2x - 10)$

$3x(x - 5) + 2(x - 5)$

$(3x + 2)(x - 5)$

EXAMPLE 2 SIMPLIFYING RATIONAL EXPRESSIONS

Simplify the following.

a. $\dfrac{x^2 + 8x + 12}{x^2 + 7x + 10}$

b. $\dfrac{6x^2 - 29x + 28}{15x^2 - 41x + 28}$

c. $\dfrac{(x + 4)(x - 3)}{(x - 3) + 5}$

d. $\dfrac{18x^2 - 3x - 10}{x^2 - 2x - 3}$

Solution

a. $\dfrac{x^2 + 8x + 12}{x^2 + 7x + 10} = \dfrac{(x + 6)(x + 2)}{(x + 5)(x + 2)}$ Factor the numerator and denominator.

$\qquad\qquad\quad = \dfrac{(x + 6)\cancel{(x + 2)}}{(x + 5)\cancel{(x + 2)}}$ Cancel any common factors.

$\qquad\qquad\quad = \dfrac{x + 6}{x + 5}$

b. $\dfrac{6x^2 - 29x + 28}{15x^2 - 41x + 28} = \dfrac{(2x - 7)(3x - 4)}{(3x - 4)(5x - 7)}$ Factor.

$\qquad\qquad\quad\quad = \dfrac{(2x - 7)\cancel{(3x - 4)}}{\cancel{(3x - 4)}(5x - 7)}$ Cancel any common factors.

$\qquad\qquad\quad\quad = \dfrac{2x - 7}{5x - 7}$

c. $\dfrac{(x + 4)(x - 3)}{(x - 3) + 5} = \dfrac{(x + 4)(x - 3)}{x + 2}$ The $x - 3$ cannot cancel because of the addition present in the denominator.

d. $\dfrac{18x^2 - 3x - 10}{x^2 - 2x - 3} = \dfrac{(3x + 2)(6x - 5)}{(x - 3)(x + 1)}$ There are no common factors, so the expression is already simplified.

In parts c and d of Example 2 you should note that the lack of common factors keeps us from simplifying the rational expression further. In all of these expressions you can multiply the factors back together as you would with numeric fractions, or you can leave them in factored form.

EXAMPLE 2 PRACTICE PROBLEM

Simplify the following.

a. $\dfrac{(x + 5)(x + 7)}{(x + 7)(x - 9)}$

b. $\dfrac{x^2 - 5x + 6}{x^2 - 13x + 30}$

c. $\dfrac{x^3 - 5x^2 + 6x}{x^2 + 5x - 14}$

d. $\dfrac{5(x + 7)}{10(x + 2) + 3}$

One special case that can occur is when an expression in the numerator is almost the same as an expression in the denominator. Most of the time the expression cannot be canceled. Sometimes, you will be able to take out a common factor and make the two expressions the same. The common factor that most often needs to be taken out is -1.

EXAMPLE **3** SPECIAL CASES FOR SIMPLIFYING

Simplify the following.

a. $\dfrac{7(x-5)}{(5-x)}$ **b.** $\dfrac{2m+6}{5(m+3)}$

Solution

a. $\dfrac{7(x-5)}{(5-x)} = \dfrac{7(x-5)}{-1(x-5)}$

$= \dfrac{7\cancel{(x-5)}}{-1\cancel{(x-5)}}$

$= -\dfrac{7}{1} = -7$

> Factor -1 out of the denominator to make the expressions the same.
>
> Cancel out common factors.

b. $\dfrac{2m+6}{5(m+3)} = \dfrac{2(m+3)}{5(m+3)}$

$= \dfrac{2\cancel{(m+3)}}{5\cancel{(m+3)}}$

$= \dfrac{2}{5}$

> The negative is typically put in front of the fraction or in the numerator.
>
> Factor completely.
>
> Cancel out common factors.

EXAMPLE **3** PRACTICE PROBLEM

Simplify the following.

a. $\dfrac{4(9-x)}{7(x-9)}$ **b.** $\dfrac{x-7}{21-3x}$

When dividing polynomials by a monomial we can divide each term of the numerator by the monomial and simplify.

EXAMPLE **4** DIVIDING A POLYNOMIAL BY A MONOMIAL

Simplify

a. $\dfrac{5x^3 + 20x^2 - 8x}{5x^2}$ **b.** $\dfrac{x^5y + 4x^4y^3 - 15x^2y^4}{x^2y}$

Solution

a. We will divide each term separately by the monomial in the denominator.

$\dfrac{5x^3 + 20x^2 - 8x}{5x^2} = \dfrac{5x^3}{5x^2} + \dfrac{20x^2}{5x^2} - \dfrac{8x}{5x^2}$ Separate each term and simplify.

$= x + 4 - \dfrac{8}{5x}$

b. We will divide each term separately by the monomial in the denominator.

$$\frac{x^5y + 4x^4y^3 - 15x^2y^4}{x^2y} = \frac{x^5y}{x^2y} + \frac{4x^4y^3}{x^2y} - \frac{15x^2y^4}{x^2y}$$

$$= x^3 + 4x^2y^2 - 15y^3$$

EXAMPLE ④ PRACTICE PROBLEM

Simplify $\dfrac{12x^6y^2 + 20x^4y^3 - 18x^2y^4}{3x^2y^3}$

LONG DIVISION OF POLYNOMIALS

When we divide a polynomial by another polynomial we can use long division. Long division of polynomials is basically the same process as long division with numbers. We will start by using long division to divide 458 by 6.

$$\begin{array}{r} 76 \\ 6\overline{)458} \\ -42 \\ \hline 38 \\ -36 \\ \hline 2 \end{array}$$

6 does not divide into 4 so we divide into 45 first. 6 divides into 45 7 times, so 7 goes above the 5 and we multiply $6(7) = 42$.

Subtract and bring down the next digit (8).

6 divides into 38 6 times, so the 6 goes above the 8 and we multiply $6(6) = 36$.

Subtract. The remainder is 2.

From this long division we get $76\frac{2}{6} = 76\frac{1}{3}$. We can check this answer by multiplying it by 6.

This same division process can be used with polynomials.

EXAMPLE ⑤ DIVIDING A POLYNOMIAL BY A POLYNOMIAL

a. Divide $3x^2 + 17x + 20$ by $x + 4$

b. Divide $10x^2 + 7x - 19$ by $2x + 3$

Solution

a.

$$\begin{array}{r} 3x \\ x + 4\overline{)3x^2 + 17x + 20} \\ \underline{3x^2 + 12x} \\ 5x + 20 \end{array}$$

Divide the first term $3x^2$ by x. Multiply $x + 4$ by $3x$ and subtract.

Bring down the next term.

$$
\begin{array}{r}
3x + 5 \\
x + 4\overline{)3x^2 + 17x + 20} \\
\underline{3x^2 + 12x} \\
5x + 20 \\
\underline{5x + 20} \\
0
\end{array}
$$

Divide the 5x by x. Multiply $x + 4$ by 5 and subtract.

The remainder is zero.

Therefore

$$\frac{3x^2 + 17x + 20}{(x + 4)} = 3x + 5$$

b.

$$
\begin{array}{r}
5x \\
2x + 3\overline{)10x^2 + 7x - 19} \\
\underline{10x^2 + 15x} \\
-8x - 19
\end{array}
$$

Divide the first term $10x^2$ by $2x$. Multiply $2x + 3$ by $5x$ and subtract.

Bring down the next term.

$$
\begin{array}{r}
5x - 4 \\
2x + 3\overline{)10x^2 + 7x - 19} \\
\underline{10x^2 + 15x} \\
-8x - 19 \\
\underline{-8x - 12} \\
-7
\end{array}
$$

Divide the -8x by 2x. Multiply $2x + 3$ by -4 and subtract.

-7 is not divisible by 2x so this is the remainder.

Therefore

$$\frac{10x^2 + 7x - 19}{2x + 3} = 5x - 4 - \frac{7}{2x + 3}$$

The remainder remains over the divisor.

EXAMPLE ⑤ PRACTICE PROBLEM

a. Divide $4x^2 + 3x - 10$ by $x + 2$

b. Divide $15x^2 - 13x + 10$ by $3x - 2$

Whenever either of the polynomials is missing a term we will replace it with a place-holder that has zero as the coefficient. In the polynomial $x^2 + 5$ we will put a $0x$ in the place of the missing first degree term to get $x^2 + 0x + 5$. If more than one term is missing you can put place holders in for each missing term.

EXAMPLE 6 DIVIDING A POLYNOMIAL BY A POLYNOMIAL

a. Divide $5x^3 + 2x^2 - 7$ by $x - 1$

b. Divide $x^4 + 2x^3 + 2x^2 - 10x - 35$ by $x^2 - 5$

Solution

a. Since the dividend is missing a first degree term we will replace that term with $0x$ to allow the terms to line up while we do the division process.

$$
\begin{array}{r}
5x^2 \\
x - 1 \overline{)5x^3 + 2x^2 + 0x - 7} \\
\underline{5x^3 - 5x^2} \\
7x^2 + 0x
\end{array}
$$

Add the $0x$ term and divide x into $5x^3$.
Multiply $5x^2$ by x and subtract.
Be careful with the signs when you subtract the negative $5x^2$.
Bring down the next term.

$$
\begin{array}{r}
5x^2 + 7x \\
x - 1 \overline{)5x^3 + 2x^2 + 0x - 7} \\
\underline{5x^3 - 5x^2} \\
7x^2 + 0x \\
\underline{7x^2 - 7x} \\
7x - 7
\end{array}
$$

Divide $7x^2$ by x. Multiply x-1 by $7x$ and subtract.

Bring down the next term.

$$
\begin{array}{r}
5x^2 + 7x + 7 \\
x - 1 \overline{)5x^3 + 2x^2 + 0x - 7} \\
\underline{5x^3 - 5x^2} \\
7x^2 + 0x \\
\underline{7x^2 - 7x} \\
7x - 7 \\
\underline{7x - 7} \\
0
\end{array}
$$

Divide $7x$ by x. Multiply x-1 by 7 and subtract.

Nothing remains so we are done.

Therefore

$$
\frac{5x^3 + 2x^2 - 7}{x - 1} = 5x^2 + 7x + 7
$$

b. The divisor is missing the first degree term so we will replace it with $0x$ and divide using long division.

$$
\begin{array}{r}
x^2 \\
x^2 + 0x - 5 \overline{)x^4 + 2x^3 + 2x^2 - 10x - 35} \\
\underline{x^4 + 0x^3 - 5x^2} \\
2x^3 + 7x^2 - 10x
\end{array}
$$

$$
\begin{array}{r}
x^2 + 2x \\
x^2 + 0x - 5 \overline{)x^4 + 2x^3 + 2x^2 - 10x - 35} \\
\underline{x^4 + 0x^3 - 5x^2} \\
2x^3 + 7x^2 - 10x \\
\underline{2x^3 + 0x^2 - 10x} \\
7x^2 + 0x - 35
\end{array}
$$

$$x^2 + 0x - 5 \overline{)\begin{array}{r} x^2 + 2x + 7 \\ x^4 + 2x^3 + 2x^2 - 10x - 35 \\ \underline{x^4 + 0x^3 - 5x^2} \\ 2x^3 + 7x^2 - 10x \\ \underline{2x^3 + 0x^2 - 10x} \\ 7x^2 + 0x - 35 \\ \underline{7x^2 + 0x - 35} \\ 0 \end{array}}$$

Therefore

$$\frac{x^4 + 2x^3 + 2x^2 - 10x - 35}{x^2 - 5} = x^2 + 2x + 7$$

EXAMPLE ⑥ PRACTICE PROBLEM

Divide $3x^3 - 4x + 16$ by $x + 2$

SYNTHETIC DIVISION

Another method used to divide polynomials is synthetic division. This is a simplified method that focuses on the coefficients of each term and eliminates the focus on the variables. This method works for single variable polynomials that are being divided by a binomial of the form $x - c$.

EXAMPLE ⑦ USING SYNTHETIC DIVISION

a. Use synthetic division to divide $3x^3 - 10x^2 - 13x + 20$ by $x - 4$

b. Use synthetic division to divide $5x^4 + 2x^3 - x - 50$ by $x + 2$

Solution

a. Since the divisor is of the form $x - c$ we know c is 4. The dividend's coefficients are 3, -10, -13, and 20.

$$\begin{array}{r|rrrr} 4 & 3 & -10 & -13 & 20 \\ & & & & \\ \hline & 3 & & & \end{array}$$
Write out the coefficients in order and c.
Draw a line and bring down the first coefficient.

$$\begin{array}{r|rrrr} 4 & 3 & -10 & -13 & 20 \\ & & 12 & & \\ \hline & 3 & & & \end{array}$$
Multiply 4 by 3 and write the product under the next coefficient.

$$\begin{array}{r|rrrr} 4 & 3 & -10 & -13 & 20 \\ & & 12 & & \\ \hline & 3 & 2 & & \end{array}$$

Add the -10 and 12 and write the sum below.

$$\begin{array}{r|rrrr} 4 & 3 & -10 & -13 & 20 \\ & & 12 & 8 & \\ \hline & 3 & 2 & -5 & \end{array}$$

Multiply 4 by 2 and write the product under the next coefficient.
Add the -13 and 8 and write the sum below.

$$\begin{array}{r|rrrr} 4 & 3 & -10 & -13 & 20 \\ & & 12 & 8 & -20 \\ \hline & 3 & 2 & -5 & 0 \end{array}$$

Multiply 4 by -5 and write the product under the next coefficient.
Add the -20 and 20 and write the sum below.

We have completed the synthetic division but we still need to translate the result into a polynomial. Each number below the line is a coefficient for a term in the final answer. The first number will be the coefficient of the first term that will have degree one less than the degree of the dividend. The dividend had a degree of 3 so we will start with a second degree term. These coefficients result in the quotient.

$$3x^2 + 2x - 5$$

The zero on the end is the remainder so this answer does not have a remainder.

b.　The divisor $x + 2$ can be written in the form $x - c$ as $x - (-2)$, so $c = -2$. The dividend is missing a second degree term so we will write zero as its coefficient. Start by writing c and the coefficients down.

$$\begin{array}{r|rrrrr} -2 & 5 & 2 & 0 & -1 & -50 \\ & & -10 & 16 & -32 & 66 \\ \hline & 5 & -8 & 16 & -33 & 16 \end{array}$$

The last number in the bottom row is 16 so that is our remainder. Therefore we have the following quotient.

$$5x^3 - 8x^2 + 16x - 33 + \frac{16}{x + 2}$$

EXAMPLE PRACTICE PROBLEM

a.　Use synthetic division to divide $-2x^3 + 4x^2 - 9x + 45$ by $x - 3$.

b.　Use synthetic division to divide $3x^5 + 4x^4 + x^2 + 6x + 600$ by $x + 5$

7.2 Exercises

In Exercises 1 through 30, simplify the rational expression.

1. $\dfrac{20x^3}{14x}$

2. $\dfrac{36x^2y}{15xy}$

3. $\dfrac{12(x+9)}{(x+5)(x+9)}$

4. $\dfrac{x+5}{(x-7)(x+5)}$

5. $\dfrac{x+3}{2x+6}$

6. $\dfrac{x-7}{3x-21}$

7. $\dfrac{8-x}{2x-16}$

8. $\dfrac{5x-20}{8-2x}$

9. $\dfrac{(x+5)(x-3)}{(x-3)(x-2)}$

10. $\dfrac{(x+4)(x+6)}{(x+6)(x-3)}$

11. $\dfrac{x^2+6x+9}{(x+2)(x+3)}$

12. $\dfrac{p-8}{p^2-12p+32}$

13. $\dfrac{t^2+2t-15}{t^2-9t+18}$

14. $\dfrac{x^2+5x-14}{x^2-6x+8}$

15. $\dfrac{w^2-16}{w^2+w-12}$

16. $\dfrac{t^2+7t-8}{t^2-64}$

17. $\dfrac{5(x+7)}{(x+7)+10}$

18. $\dfrac{(x-5)}{2(x-5)+3}$

19. $\dfrac{m+3}{2(m+3)-5}$

20. $\dfrac{4(k-6)-7}{5(k-6)}$

21. $\dfrac{2x^2-23x-70}{x^2-21x+98}$

22. $\dfrac{2x^2+11x+15}{4x^2+17x+15}$

23. $\dfrac{10x^3+5x^2-4x}{5x}$

24. $\dfrac{15t^4+20t^2+6t}{5t}$

25. $\dfrac{3x^4-8x^3+x^2}{4x^3}$

26. $\dfrac{5b^3-4b^2-6b}{4b^2}$

27. $\dfrac{10a^3b^2+12a^2b^3-14ab^2}{2ab}$

28. $\dfrac{24m^4n^3-16m^2n+10mn^2}{8mn}$

29. $\dfrac{10g^5h^3-30g^4h^2+20g^2h}{5g^2h^2}$

30. $\dfrac{x^4y^2-2x^3y^4+5xy^3}{2x^2y^2}$

In Exercises 31 through 38, divide using long division.

31. $(4x^2+17x+15)\div(x+3)$

32. $(2x^2+11x+15)\div(x+3)$

33. $(3x^2+17x+20)\div(x+4)$

34. $(7x^2+38x+15)\div(x+5)$

35. $(x^2-3x-28)\div(x+4)$

36. $(t^2+7t-18)\div(t+9)$

37. $(2m^2 - 5m - 42) \div (m - 6)$

38. $(3g^2 - 19g + 20) \div (g - 5)$

In Exercises 39 through 54, divide using long division or synthetic division.

39. $(12x^2 + 32x + 21) \div (2x + 3)$

40. $(12x^2 + 23x + 10) \div (3x + 2)$

41. $(2x^3 + 11x^2 + 17x + 20) \div (x + 4)$

42. $(3x^3 + 14x^2 + 19x + 12) \div (x + 3)$

43. $(x^4 + 2x^3 - 10x - 25) \div (x^2 - 5)$

44. $(b^4 + 6b^3 - 18b - 9) \div (b^2 - 3)$

45. $(x^4 - 3x^3 + 8x^2 - 12x + 16) \div (x^2 + 4)$

46. $(x^4 - 7x^3 + 10x^2 - 35x + 25) \div (x^2 + 5)$

47. $(t^3 + 3t^2 - 12t + 16) \div (t + 3)$

48. $(m^3 - 4m^2 - 8m + 32) \div (m - 4)$

49. $(a^4 + 2a^3 + 8a^2 + 10a + 15) \div (a^2 + 5)$

50. $(w^4 - 5w^3 + w^2 + 30w - 42) \div (w^2 - 6)$

51. $(t^3 + 4t^2 - 24) \div (t - 2)$

52. $(x^3 + 5x^2 - 18) \div (x + 3)$

53. $(4a^3 + 2a + 36) \div (a + 2)$

54. $(6x^3 + 4x - 10) \div (x - 1)$

Multiplying and Dividing Rational Expressions

In this section we will study how to multiply and divide rational expressions. In the next section we will study how to add and subtract rational expressions. You should notice in these two sections that working with rational expressions is very much the same as working with fractions. The connection between working with rational expressions and working with fractions is important and can be a good basis for understanding these two sections of this chapter.

Simplifying is a key part of all operations involving rational expressions. When multiplying or dividing, you can simplify before the multiplication is performed or after. Many students find it easier to simplify the rational expressions first so that there is less multiplication to be done. Remember to always check the final product for any further simplification that can be done.

MULTIPLYING RATIONAL EXPRESSIONS

When multiplying or dividing fractions, you do not need common denominators. This makes it easier to multiply or divide any two rational expressions. Recall that when multiplying fractions, you simply multiply the numerators together and multiply the denominators together. Don't forget to simplify your results.

MULTIPLYING RATIONAL EXPRESSIONS

1. Factor the numerator and denominator of each fraction (if needed).
2. Cancel any common factors.
3. Multiply the numerators together and multiply the denominators together.
4. Leave in factored form.

EXAMPLE 1   MULTIPLYING RATIONAL EXPRESSIONS

Multiply the following.

a. $\dfrac{3}{5} \cdot \dfrac{2}{7}$

b. $\dfrac{20x}{18y^2} \cdot \dfrac{6xy}{10}$

c. $\dfrac{x+5}{x-3} \cdot \dfrac{x+2}{x+7}$

Solution

a. $\dfrac{3}{5} \cdot \dfrac{2}{7} = \dfrac{6}{35}$ There are no common factors to cancel out, so multiply the numerators and denominators together.

b. $\dfrac{20x}{18y^2} \cdot \dfrac{6xy}{10} = \dfrac{2 \cdot 2 \cdot 5 \cdot x}{2 \cdot 3 \cdot 3 \cdot y \cdot y} \cdot \dfrac{2 \cdot 3 \cdot x \cdot y}{2 \cdot 5}$ Factor the numerators and denominators.

$\qquad\qquad = \dfrac{\cancel{2} \cdot 2 \cdot \cancel{5} \cdot x}{\cancel{2} \cdot \cancel{3} \cdot 3 \cdot \cancel{y} \cdot y} \cdot \dfrac{2 \cdot \cancel{3} \cdot x \cdot \cancel{y}}{\cancel{2} \cdot \cancel{5}}$ Cancel out common factors.

$\qquad\qquad = \dfrac{2x^2}{3y}$

c. $\dfrac{x+5}{x-3} \cdot \dfrac{x+2}{x+7} = \dfrac{(x+5)(x+2)}{(x-3)(x+7)}$ There are no common factors, so multiply together.

EXAMPLE ① PRACTICE PROBLEM

Multiply the following.

a. $\dfrac{50a^2b}{20c^2} \cdot \dfrac{6ac}{9b}$

b. $\dfrac{x+5}{x-3} \cdot \dfrac{x+7}{x+5}$

If the rational expressions are more complicated make sure that you focus on one part at a time so that you do not forget any piece of the expressions you are multiplying.

EXAMPLE 2 MULTIPLYING RATIONAL EXPRESSIONS

Multiply the following.

a. $\dfrac{(x+3)(x-4)}{(x-5)(x-4)} \cdot \dfrac{(x+2)(x+7)}{(x+3)(x+7)}$

b. $\dfrac{x^2-3x-18}{x^2+7x+10} \cdot \dfrac{x^2+3x-10}{x^2+7x+12}$

Solution

a. $\dfrac{(x+3)(x-4)}{(x-5)(x-4)} \cdot \dfrac{(x+2)(x+7)}{(x+3)(x+7)} = \dfrac{\cancel{(x+3)}\cancel{(x-4)}}{(x-5)\cancel{(x-4)}} \cdot \dfrac{(x+2)\cancel{(x+7)}}{\cancel{(x+3)}\cancel{(x+7)}}$

$\qquad\qquad = \dfrac{x+2}{x-5}$ Cancel common factors and multiply.

b. $\dfrac{x^2-3x-18}{x^2+7x+10} \cdot \dfrac{x^2+3x-10}{x^2+7x+12} = \dfrac{(x+3)(x-6)}{(x+5)(x+2)} \cdot \dfrac{(x+5)(x-2)}{(x+3)(x+4)}$ Factor.

$\qquad\qquad = \dfrac{\cancel{(x+3)}(x-6)}{\cancel{(x+5)}(x+2)} \cdot \dfrac{\cancel{(x+5)}(x-2)}{\cancel{(x+3)}(x+4)}$ Reduce.

$\qquad\qquad = \dfrac{(x-6)(x-2)}{(x+2)(x+4)}$ Multiply.

EXAMPLE (2) PRACTICE PROBLEM

Multiply the following.

a. $\dfrac{(x + 3)(x + 4)}{(x - 7)(x + 3)} \cdot \dfrac{(x - 2)(x - 5)}{(x + 4)(x - 6)}$

b. $\dfrac{x^2 - 5x - 14}{2x^2 - x - 15} \cdot \dfrac{x^2 + 2x - 15}{3x^2 + 4x - 4}$

DIVIDING RATIONAL EXPRESSIONS

Dividing rational expressions is the same as dividing numerical fractions in that you multiply by the reciprocal of the fraction you are dividing by. This means that division will simply be the same as multiplication after you have flipped over the second fraction.

DIVIDING RATIONAL EXPRESSIONS

1. Multiply by the reciprocal of the second fraction. (Flip over the fraction you are dividing by and make the division into multiplication.)

2. Factor the numerator and denominator of each fraction (if needed).

3. Cancel any common factors.

4. Multiply the numerators together and multiply the denominators together.

5. Leave in factored form or multiply remaining factors together.

EXAMPLE 3 DIVIDING RATIONAL EXPRESSIONS

Divide the following.

a. $\dfrac{2}{3} \div \dfrac{5}{7}$

b. $\dfrac{\frac{7}{10}}{\frac{2}{5}}$

c. $\dfrac{x + 3}{x - 7} \div \dfrac{x - 4}{x - 7}$

Solution

a. $\dfrac{2}{3} \div \dfrac{5}{7} = \dfrac{2}{3} \cdot \dfrac{7}{5}$ **Multiply by the reciprocal of the second fraction.**

$= \dfrac{14}{15}$

b. $\dfrac{\dfrac{7}{10}}{\dfrac{2}{5}} = \dfrac{7}{10} \cdot \dfrac{5}{2}$ Multiply by the reciprocal of the fraction you are dividing by.

$\quad = \dfrac{7}{2 \cdot \cancel{5}} \cdot \dfrac{\cancel{5}}{2}$ Factor and cancel common factors.

$\quad = \dfrac{7}{4}$

$\quad = 1\dfrac{3}{4}$ Reduce.

c. $\dfrac{x+3}{x-7} \div \dfrac{x-4}{x-7} = \dfrac{x+3}{x-7} \cdot \dfrac{x-7}{x-4}$

$\quad = \dfrac{x+3}{\cancel{x-7}} \cdot \dfrac{\cancel{x-7}}{x-4}$ Cancel common factors.

$\quad = \dfrac{x+3}{x-4}$ Multiply.

EXAMPLE ③ PRACTICE PROBLEM

Divide the following.

a. $\dfrac{4x^3}{10y^2} \div \dfrac{6x}{15y^3}$

b. $\dfrac{x+5}{x-9} \div \dfrac{x+5}{x+7}$

EXAMPLE 4 DIVIDING RATIONAL EXPRESSIONS

Divide the following.

a. $\dfrac{(x+2)(x+5)}{(x-3)(2x+7)} \div \dfrac{(x-3)(x+5)}{(2x+7)(x-9)}$

b. $\dfrac{2x^2+11x+12}{x^2-11x+30} \div \dfrac{2x^2+15x+18}{x^2+2x-35}$

Solution

a. $\dfrac{(x+2)(x+5)}{(x-3)(2x+7)} \div \dfrac{(x-3)(x+5)}{(2x+7)(x-9)} = \dfrac{(x+2)(x+5)}{(x-3)(2x+7)} \cdot \dfrac{(2x+7)(x-9)}{(x-3)(x+5)}$

$\quad = \dfrac{(x+2)\cancel{(x+5)}}{(x-3)\cancel{(2x+7)}} \cdot \dfrac{\cancel{(2x+7)}(x-9)}{(x-3)\cancel{(x+5)}}$

$\quad = \dfrac{(x+2)(x-9)}{(x-3)(x-3)}$ Multiply by the reciprocal then cancel common factors.

$\quad = \dfrac{(x+2)(x-9)}{(x-3)^2}$

b. $\dfrac{2x^2 + 11x + 12}{x^2 - 11x + 30} \div \dfrac{2x^2 + 15x + 18}{x^2 + 2x - 35} = \dfrac{2x^2 + 11x + 12}{x^2 - 11x + 30} \cdot \dfrac{x^2 + 2x - 35}{2x^2 + 15x + 18}$

$$= \dfrac{(x + 4)(2x + 3)}{(x - 5)(x - 6)} \cdot \dfrac{(x - 5)(x + 7)}{(2x + 3)(x + 6)}$$

$$= \dfrac{(x + 4)\cancel{(2x + 3)}}{\cancel{(x - 5)}(x - 6)} \cdot \dfrac{\cancel{(x - 5)}(x + 7)}{\cancel{(2x + 3)}(x + 6)}$$

$$= \dfrac{(x + 4)(x + 7)}{(x - 6)(x + 6)}$$

EXAMPLE 4 PRACTICE PROBLEM

Divide the following.

a. $\dfrac{(x + 2)(x + 4)}{(x - 3)(x + 7)} \div \dfrac{(x + 4)(x - 8)}{(x + 7)(x - 5)}$

b. $\dfrac{x^2 - 4x - 21}{2x^2 - 9x - 35} \div \dfrac{x^2 + 12x + 27}{2x^2 - 3x - 20}$

7.3 Exercises

For Exercises 1 through 16, multiply the rational expressions:

1. $\dfrac{15x^2}{9y^3} \cdot \dfrac{21y^5}{35x}$

2. $\dfrac{a^3b^4}{10c} \cdot \dfrac{4ac}{b}$

3. $\dfrac{35a^2}{12b^3c} \cdot \dfrac{40b}{a^5c}$

4. $\dfrac{3xy}{8z^2} \cdot \dfrac{6z^4}{5x}$

5. $\dfrac{x + 3}{x + 7} \cdot \dfrac{x + 7}{x - 5}$

6. $\dfrac{x - 8}{x - 4} \cdot \dfrac{x - 5}{x - 8}$

7. $\dfrac{2(x - 4)}{3x - 12} \cdot \dfrac{3(x + 5)}{x - 7}$

8. $\dfrac{5(w + 7)}{2w + 10} \cdot \dfrac{4(w + 5)}{w - 11}$

9. $\dfrac{(x + 3)(x + 7)}{(x - 2)(x + 3)} \cdot \dfrac{(x + 7)(x - 2)}{(x - 3)(x - 9)}$

10. $\dfrac{(c - 3)(c + 6)}{(c - 7)(c + 2)} \cdot \dfrac{(c - 7)(c - 11)}{(c + 6)(c - 3)}$

11. $\dfrac{(k + 5)(7 - k)}{(k - 7)(k - 3)} \cdot \dfrac{(k - 3)(k + 6)}{(k + 9)(k + 5)}$

12. $\dfrac{(t + 7)(t - 9)}{(t + 3)(t + 7)} \cdot \dfrac{(t + 3)(t + 13)}{(9 - t)(t - 13)}$

13. $\dfrac{m^2 + 8m + 7}{m^2 - 2m - 3} \cdot \dfrac{m^2 - 9}{m^2 + 9m + 14}$

14. $\dfrac{t^2 - 16}{t^2 + 9t + 14} \cdot \dfrac{t^2 + 4t - 21}{t^2 - t - 20}$

15. $\dfrac{x^2 - 16x + 55}{x^2 - x - 12} \cdot \dfrac{x^2 + 12x + 27}{x^2 - 9x + 20}$

16. $\dfrac{m^2 + 6m + 9}{m^2 + 5m + 6} \cdot \dfrac{m^2 - 6m - 16}{m^2 - 9}$

In Exercises 17 through 32, divide the rational expressions

17. $\dfrac{x^2 y}{z^3} \div \dfrac{x}{z}$

18. $\dfrac{10a^3}{6b^2c^4} \div \dfrac{15a}{20bc^2}$

19. $\dfrac{x+5}{x-3} \div \dfrac{x-5}{x-7}$

20. $\dfrac{x+15}{x-3} \div \dfrac{x+7}{x-3}$

21. $\dfrac{5x+20}{x-4} \div \dfrac{3x+12}{x+7}$

22. $\dfrac{w+2}{10w+15} \div \dfrac{7w+14}{2w+3}$

23. $\dfrac{(x+3)(x+2)}{(x-8)(x-7)} \div \dfrac{(x+2)(x-5)}{(x-7)(x-5)}$

24. $\dfrac{(m+7)(m-5)}{(m+2)(m+7)} \div \dfrac{(m-5)(m-12)}{(m+2)(m+6)}$

25. $\dfrac{w^2+8w+15}{w^2+12w+35} \div \dfrac{w^2-5w-24}{w^2+3w-28}$

26. $\dfrac{k^2+k-20}{k^2+10k+21} \div \dfrac{k^2-10k+24}{k^2-13k+42}$

27. $\dfrac{c^2-c-20}{c^2+6c+8} \div \dfrac{2c^2+11c+12}{3c^2+c-10}$

28. $\dfrac{3k^2+19k+20}{k^2+6k+5} \div \dfrac{2k^2+2k-40}{k^2-2k-3}$

29. $\dfrac{6t^2+t-35}{10t^2+17t-20} \div \dfrac{12t^2-t-63}{20t^2+29t-36}$

30. $\dfrac{2x^2-5x-12}{x^2-3x-28} \div \dfrac{6x^2-7x-24}{3x^2+13x-56}$

31. $\dfrac{x^2-25}{x^2+7x+10} \div \dfrac{x^2-4}{x^2+9x+8}$

32. $\dfrac{t^2+12t+20}{t^2-49} \div \dfrac{t^2-100}{t^2-9t+14}$

SECTION
7.4

Adding and Subtracting Rational Expressions

Now that we have studied simplifying, multiplying, and dividing rational expressions, we are going to learn how to add and subtract them. Adding and subtracting fractions requires us to have common denominators. This has always been a condition for adding or subtracting fractions and requires a little more work than multiplying and dividing, in which we do not need common denominators. Being able to add or subtract rational expressions will allow us to simplify situations in which several rational expressions are used.

LEAST COMMON DENOMINATOR

Finding a common denominator for a rational expression will be the same process as finding one for a numeric fraction, but once again we will need to focus more carefully on each step of the process. When finding the least common denominator (LCD) of a fraction, you need to factor each denominator and compare them so that you can choose the right combination of factors that will make up the least common denominator. In some cases you might be able to figure out the LCD without doing all of these steps, but with rational expressions it is going to take a little more patience.

FINDING THE LEAST COMMON DENOMINATOR FOR RATIONAL EXPRESSIONS

1. Factor the denominators (if needed).
2. Take the highest power of each factor for the LCD.
3. Leave in factored form or multiply the factors together.
   We generally leave polynomials factored but multiply out single terms.

WRITING FRACTIONS IN TERMS OF THE LCD

1. Find the least common denominator.
2. Determine what factors of the LCD the fraction's denominator is missing.
3. Multiply the numerator and denominator of the fraction by the missing factors.

EXAMPLE ☐ 1 **FINDING THE LCD WITH MONOMIAL DENOMINATORS**

Find the least common denominator for each of the following sets of fractions and write each fraction in terms of the LCD.

a. $\dfrac{12}{35}$ $\dfrac{8}{165}$ **b.** $\dfrac{2}{5x}$ $\dfrac{3}{10x^2}$ **c.** $\dfrac{6ab}{25c^3d}$ $\dfrac{a}{14bcd^2}$

Solution

a.
$$\frac{12}{35} \quad \frac{8}{165}$$

$$\frac{12}{5 \cdot 7} \qquad \frac{8}{3 \cdot 5 \cdot 11} \qquad \text{Factor both denominators.}$$

$$LCD = 3 \cdot 5 \cdot 7 \cdot 11 \qquad \text{Take a copy of each factor for the LCD.}$$

$$LCD = 1155$$

$$\frac{3 \cdot 11}{3 \cdot 11} \cdot \frac{12}{5 \cdot 7} \qquad \frac{8}{3 \cdot 5 \cdot 11} \cdot \frac{7}{7} \qquad \text{Multiply each fraction by the factors that its denominator is missing.}$$

$$\frac{396}{1155} \qquad \frac{56}{1155}$$

b.
$$\frac{2}{5x} \quad \frac{3}{10x^2}$$

$$\frac{2}{5 \cdot x} \qquad \frac{3}{2 \cdot 5 \cdot x^2} \qquad \text{Factor both denominators.}$$

$$LCD = 2 \cdot 5 \cdot x^2 \qquad \text{Take the highest power of each}$$

$$LCD = 10x^2 \qquad \text{factor for the LCD.}$$

$$\frac{2 \cdot x}{2 \cdot x} \cdot \frac{2}{5 \cdot x} \qquad \frac{3}{2 \cdot 5 \cdot x^2} \qquad \text{Multiply each fraction by the factors that its denominator is missing.}$$

$$\frac{4x}{10x^2} \qquad \frac{3}{10x^2}$$

c.

$$\frac{6ab}{25c^3d} \qquad \frac{a}{14bcd^2}$$

$$\frac{2 \cdot 3 \cdot a \cdot b}{5^2 \cdot c^3 \cdot d} \qquad \frac{a}{2 \cdot 7 \cdot b \cdot c \cdot d^2} \qquad \text{Factor both denominators.}$$

$$\text{LCD} = 2 \cdot 5^2 \cdot 7 \cdot b \cdot c^3 \cdot d^2 \qquad \begin{array}{l}\text{Take the highest power of each}\\ \text{factor for the LCD.}\end{array}$$

$$\text{LCD} = 700bc^3d^2$$

$$\frac{2 \cdot 7 \cdot b \cdot d}{2 \cdot 7 \cdot b \cdot d} \cdot \frac{2 \cdot 3 \cdot a \cdot b}{5^2 \cdot c^3 \cdot d} \qquad \frac{a}{2 \cdot 7 \cdot b \cdot c \cdot d^2} \cdot \frac{5^2 \cdot c^2}{5^2 \cdot c^2} \qquad \begin{array}{l}\text{Multiply each fraction}\\ \text{by the factors that its}\\ \text{denominator is}\\ \text{missing.}\end{array}$$

$$\frac{84ab^2d}{700bc^3d^2} \qquad \frac{25ac^2}{700bc^3d^2}$$

EXAMPLE ① PRACTICE PROBLEM

Find the least common denominator for the following two fractions and write each fraction in terms of the LCD.

$$\frac{7}{24m} \qquad \frac{11}{45m^3n}$$

With simple denominators it is easier to find the factors you need to make up the LCD. Note that you want to take the highest power of each factor that is present to be a part of the LCD. When you are working with more complicated denominators, you will find situations in which the denominators look very similar but cannot be changed easily. Let's find the LCD for the two fractions.

$$\frac{1}{x+2} \qquad \frac{1}{x+4}$$

It seems as though you could either add 2 to $x + 2$ or multiply $x + 2$ by 2 and get $x + 4$. The problem with these ideas is that multiplying by 2 will give you $2x + 4$, not $x + 4$, and adding 2 to the denominator is not possible because that would change the value of the fraction you are working on. The only way to get a LCD from these two denominators is to use both of them as factors in the new denominator giving you $(x + 4)(x + 2)$ as the LCD.

$$\frac{(x+4)}{(x+4)} \cdot \frac{1}{(x+2)} \qquad \frac{1}{(x+4)} \cdot \frac{(x+2)}{(x+2)}$$

$$\frac{x+4}{(x+4)(x+2)} \qquad \frac{x+2}{(x+4)(x+2)}$$

EXAMPLE ② FINDING THE LCD WITH POLYNOMIAL DENOMINATORS

Find the least common denominator for each of the following sets of fractions and write each fraction in terms of the LCD.

a. $\dfrac{x+5}{x-6} \qquad \dfrac{x+2}{x+7}$

b. $\dfrac{x+3}{(x-5)(x+4)} \qquad \dfrac{x-7}{(x-5)(x+2)}$

c. $\dfrac{5x + 7}{x^2 + 5x + 6} \qquad \dfrac{3x - 8}{2x^2 + 9x + 9}$

Solution

a. $\dfrac{x + 5}{x - 6} \qquad \dfrac{x + 2}{x + 7}$ The denominators are already factored, so we take a copy of each factor for the LCD.

$LCD = (x - 6)(x + 7)$

$LCD = x^2 + x - 42$

$\dfrac{(x + 7)}{(x + 7)} \cdot \dfrac{(x + 5)}{(x - 6)} \qquad \dfrac{(x + 2)}{(x + 7)} \cdot \dfrac{(x - 6)}{(x - 6)}$

$\dfrac{(x + 7)(x + 5)}{(x - 6)(x + 7)} \qquad \dfrac{(x + 2)(x - 6)}{(x - 6)(x + 7)}$ We will leave the fractions in this factored form.

b. $\dfrac{x + 3}{(x - 5)(x + 4)} \qquad \dfrac{x - 7}{(x - 5)(x + 2)}$ Because each denominator has only one copy of $(x - 5)$, we will only need one copy in the LCD.

$\qquad\qquad LCD = (x - 5)(x + 4)(x + 2)$

$\qquad\qquad LCD = x^3 + x^2 - 22x - 40$

$\dfrac{(x + 2)}{(x + 2)} \cdot \dfrac{(x + 3)}{(x - 5)(x + 4)} \qquad \dfrac{(x - 7)}{(x - 5)(x + 2)} \cdot \dfrac{(x + 4)}{(x + 4)}$

$\dfrac{(x + 2)(x + 3)}{(x - 5)(x + 4)(x + 2)} \qquad \dfrac{(x - 7)(x + 4)}{(x - 5)(x + 4)(x + 2)}$

c. $\dfrac{5x + 7}{x^2 + 5x + 6} \qquad \dfrac{3x - 8}{2x^2 + 9x + 9}$ Factor the denominators.

$\dfrac{5x + 7}{(x + 3)(x + 2)} \qquad \dfrac{3x - 8}{(2x + 3)(x + 3)}$

$\qquad\qquad LCD = (x + 3)(x + 2)(2x + 3)$

$\qquad\qquad LCD = 2x^3 + 13x^2 + 27x + 18$

$\dfrac{(2x + 3)}{(2x + 3)} \cdot \dfrac{(5x + 7)}{(x + 3)(x + 2)} \qquad \dfrac{(3x - 8)}{(2x + 3)(x + 3)} \cdot \dfrac{(x + 2)}{(x + 2)}$

$\dfrac{(2x + 3)(5x + 7)}{(x + 3)(x + 2)(2x + 3)} \qquad \dfrac{(3x - 8)(x + 2)}{(x + 3)(x + 2)(2x + 3)}$

EXAMPLE ② PRACTICE PROBLEM

Find the least common denominator for each of the following sets of fractions and write each fraction in terms of the LCD.

a. $\dfrac{x + 3}{x + 8} \qquad \dfrac{x + 5}{x + 9}$

b. $\dfrac{x + 2}{(x - 4)(x - 7)} \qquad \dfrac{x + 4}{(x - 7)(x + 8)}$

c. $\dfrac{3x + 2}{6x^2 - 7x - 20} \qquad \dfrac{4x - 9}{2x^2 - 19x + 35}$

ADDING RATIONAL EXPRESSIONS

Getting the common denominator is the hardest part of adding and subtracting rational expressions. Once you have the common denominator, you can just add or subtract the numerators, and the LCD remains the denominator of the sum.

ADDING RATIONAL EXPRESSIONS

1. Write all fractions with a common denominator.

2. Add the numerators and keep the denominator the same.

3. Factor the numerator and reduce (if possible).

EXAMPLE　**3**　ADDING SIMPLE RATIONAL EXPRESSIONS

Add the following rational expressions.

a. $\dfrac{5}{7x^2} + \dfrac{12}{5x}$

b. $\dfrac{x+2}{x-3} + \dfrac{x-7}{x-3}$

c. $x + \dfrac{7}{x+3}$

Solution

a. $\dfrac{5}{7x^2} + \dfrac{12}{5x} = \dfrac{5}{5} \cdot \dfrac{5}{7x^2} + \dfrac{12}{5x} \cdot \dfrac{7x}{7x}$ 　　Get a common denominator and rewrite each fraction over the LCD.

$\qquad = \dfrac{25}{35x^2} + \dfrac{84x}{35x^2}$ 　　Add the numerators. In this case they are not like terms, so you can write them with addition between them in one fraction.

$\qquad = \dfrac{84x + 25}{35x^2}$

b. $\dfrac{x+2}{x-3} + \dfrac{x-7}{x-3} = \dfrac{2x-5}{x-3}$ 　　Because the denominators are already the same, we can simply add the numerators together.

c. $x + \dfrac{7}{x+3} = \dfrac{x}{1} + \dfrac{7}{x+3}$ 　　The x has a denominator of 1, so we will need to get a common denominator.

$\qquad = \dfrac{(x+3)}{(x+3)} \cdot \dfrac{x}{1} + \dfrac{7}{x+3}$

$\qquad = \dfrac{(x+3)x}{x+3} + \dfrac{7}{x+3}$

$\qquad = \dfrac{x^2 + 3x}{x+3} + \dfrac{7}{x+3}$ 　　We need to multiply out the numerators, so we can add them together.

$\qquad = \dfrac{x^2 + 3x + 7}{x+3}$

EXAMPLE ③ PRACTICE PROBLEM

Add the following rational expressions.

a. $\dfrac{16}{5x^3} + \dfrac{7}{2x}$

b. $x + \dfrac{4}{x-5}$

EXAMPLE ④ ADDING RATIONAL EXPRESSIONS

Add the following rational expressions.

a. $\dfrac{x+2}{(x+3)(x+5)} + \dfrac{x-7}{x+5}$

b. $\dfrac{x+5}{x^2+5x-24} + \dfrac{x-9}{x^2+12x+32}$

Solution

a. $\dfrac{x+2}{(x+3)(x+5)} + \dfrac{x-7}{x+5}$

$= \dfrac{x+2}{(x+3)(x+5)} + \dfrac{x-7}{x+5} \cdot \dfrac{(x+3)}{(x+3)}$ **Get a common denominator.**

$= \dfrac{x+2}{(x+3)(x+5)} + \dfrac{(x-7)(x+3)}{(x+5)(x+3)}$

$= \dfrac{x+2}{(x+3)(x+5)} + \dfrac{x^2-4x-21}{(x+3)(x+5)}$ **Multiply the numerator out.**

$= \dfrac{x^2-3x-19}{(x+3)(x+5)}$ **Add the numerators.**

b. $\dfrac{x+5}{x^2+5x-24} + \dfrac{x-9}{x^2+12x+32} = \dfrac{x+5}{(x-3)(x+8)} + \dfrac{x-9}{(x+8)(x+4)}$

$= \dfrac{(x+4)}{(x+4)} \cdot \dfrac{x+5}{(x-3)(x+8)} + \dfrac{x-9}{(x+8)(x+4)} \cdot \dfrac{(x-3)}{(x-3)}$

$= \dfrac{(x+4)(x+5)}{(x-3)(x+8)(x+4)} + \dfrac{(x-9)(x-3)}{(x-3)(x+8)(x+4)}$

$= \dfrac{x^2+9x+20}{(x-3)(x+8)(x+4)} + \dfrac{x^2-12x+27}{(x-3)(x+8)(x+4)}$

$= \dfrac{2x^2-3x+47}{(x-3)(x+8)(x+4)}$

EXAMPLE ④ PRACTICE PROBLEM

Add the following rational expressions.

a. $\dfrac{2x+5}{(x+3)(x-4)} + \dfrac{3x+7}{(x+3)(x+7)}$

SUBTRACTING RATIONAL EXPRESSIONS

Subtraction is the same process as addition but with one warning: Be sure to subtract the entire numerator of the second fraction. The most common error that students make is to forget to distribute the subtraction over the numerator of the second fraction.

SUBTRACTING RATIONAL EXPRESSIONS

1. Write all fractions with a common denominator.

2. Subtract the numerators and keep the denominator the same.

3. Factor the numerator and reduce (if possible).

Caution: When subtracting rational expressions, be sure to distribute the subtraction to the entire numerator of the second fraction.

EXAMPLE **5** SUBTRACTING RATIONAL EXPRESSIONS

Subtract the following rational expressions.

a. $\dfrac{x + 5}{x + 8} - \dfrac{x - 9}{x + 8}$

b. $\dfrac{x - 7}{(x + 2)(x - 4)} - \dfrac{x + 5}{(x + 6)(x + 2)}$

c. $\dfrac{3x + 2}{x^2 - 3x - 10} - \dfrac{2x + 7}{x^2 - 2x - 15}$

Solution

a. $\dfrac{x + 5}{x + 8} - \dfrac{x - 9}{x + 8} = \dfrac{x + 5 - (x - 9)}{x + 8}$

$= \dfrac{x + 5 - x + 9}{x + 8} = \dfrac{14}{x + 8}$ **Distribute the negative and combine like terms.**

b. $\dfrac{x - 7}{(x + 2)(x - 4)} - \dfrac{x + 5}{(x + 6)(x + 2)}$

$= \dfrac{(x + 6)}{(x + 6)} \cdot \dfrac{(x - 7)}{(x + 2)(x - 4)} - \dfrac{(x + 5)}{(x + 6)(x + 2)} \cdot \dfrac{(x - 4)}{(x - 4)}$

$= \dfrac{(x + 6)(x - 7)}{(x + 6)(x + 2)(x - 4)} - \dfrac{(x + 5)(x - 4)}{(x + 6)(x + 2)(x - 4)}$

$= \dfrac{x^2 - x - 42}{(x + 6)(x + 2)(x - 4)} - \dfrac{x^2 + x - 20}{(x + 6)(x + 2)(x - 4)}$

$= \dfrac{x^2 - x - 42 - (x^2 + x - 20)}{(x + 6)(x + 2)(x - 4)}$

$= \dfrac{x^2 - x - 42 - x^2 - x + 20}{(x + 6)(x + 2)(x - 4)}$ **Distribute the negative.**

$= \dfrac{-2x - 22}{(x + 6)(x + 2)(x - 4)}$

c. $\dfrac{3x+2}{x^2-3x-10}-\dfrac{2x+7}{x^2-2x-15}=\dfrac{3x+2}{(x-5)(x+2)}-\dfrac{2x+7}{(x-5)(x+3)}$

$$=\frac{(x+3)}{(x+3)}\cdot\frac{(3x+2)}{(x-5)(x+2)}-\frac{(2x+7)}{(x-5)(x+3)}\cdot\frac{(x+2)}{(x+2)}$$

$$=\frac{3x^2+11x+6}{(x+3)(x-5)(x+2)}-\frac{2x^2+11x+14}{(x+3)(x-5)(x+2)}$$

$$=\frac{3x^2+11x+6-(2x^2+11x+14)}{(x+3)(x-5)(x+2)}$$

$$=\frac{3x^2+11x+6-2x^2-11x-14}{(x+3)(x-5)(x+2)}$$

$$=\frac{x^2-8}{(x+3)(x-5)(x+2)}$$

EXAMPLE ⑤ PRACTICE PROBLEM

Subtract the following rational expressions.

a. $\dfrac{5x-4}{(x+5)(x+6)}-\dfrac{3x-16}{(x+5)(x+6)}$

b. $\dfrac{3x-7}{x^2-4x-45}-\dfrac{2x+3}{x^2-5x-36}$

SIMPLIFYING COMPLEX FRACTIONS

In some circumstances we will have rational expressions that contain fractions in their numerator, denominator, or both. We call these types of expression complex fractions. Some examples of complex fractions are

$$\frac{\dfrac{2}{x}+5}{7x+\dfrac{3}{x+1}}\qquad\qquad\frac{\dfrac{5}{x+7}}{\dfrac{2}{x-8}}$$

Simplifying complex fractions uses many of the same skills we have learned in this section and the previous two sections. The least common denominator of the fractions plays a key role in the simplification process. Multiplying the numerator and denominator of the rational expression by the LCD of all the fractions involved will eliminate the fractions in both the numerator and denominator and simplify the rational expression.

SIMPLIFYING COMPLEX FRACTIONS

1. Find the LCD of all fractions inside the rational expression.

2. Multiply the numerator and denominator of the rational expression by the LCD.

3. Factor and reduce (if possible).

E X A M P L E **6** SIMPLIFYING COMPLEX FRACTIONS

Simplify the following complex fractions.

a.
$$\dfrac{5 + \dfrac{2}{x+7}}{\dfrac{3}{x+2}}$$

b.
$$\dfrac{\dfrac{2}{x} + \dfrac{7}{y^2}}{\dfrac{3}{x^3} - \dfrac{4}{y}}$$

Solution

a.

$$\dfrac{5 + \dfrac{2}{x+7}}{\dfrac{3}{x+2}} = \dfrac{5 + \dfrac{2}{x+7}}{\dfrac{3}{x+2}} \cdot \dfrac{(x+7)(x+2)}{(x+7)(x+2)}$$

Multiply the numerator and denominator by the LCD.

$$= \dfrac{5(x+7)(x+2) + \dfrac{2}{x+7}(x+7)(x+2)}{\dfrac{3}{x+2}(x+7)(x+2)}$$

Distribute to every term.

$$= \dfrac{5(x^2 + 9x + 14) + 2(x+2)}{3(x+7)}$$

Multiply out and combine like terms.

$$= \dfrac{5x^2 + 45x + 70 + 2x + 4}{3x + 21}$$

$$= \dfrac{5x^2 + 47x + 74}{3x + 21}$$

$$= \dfrac{(5x + 37)(x+2)}{3(x+7)}$$

Factor and simplify if possible.

b.

$$\dfrac{\dfrac{2}{x} + \dfrac{7}{y^2}}{\dfrac{3}{x^3} - \dfrac{4}{y}} = \dfrac{\dfrac{2}{x} + \dfrac{7}{y^2}}{\dfrac{3}{x^3} - \dfrac{4}{y}} \cdot \dfrac{x^3 y^2}{x^3 y^2}$$

Multiply the numerator and denominator by the LCD.

$$= \dfrac{\dfrac{2}{x}(x^3 y^2) + \dfrac{7}{y^2}(x^3 y^2)}{\dfrac{3}{x^3}(x^3 y^2) - \dfrac{4}{y}(x^3 y^2)}$$

Distribute to every term.

$$= \dfrac{2x^2 y^2 + 7x^3}{3y^2 - 4x^3 y}$$

$$= \dfrac{x^2(2y^2 + 7x)}{y(3y - 4x^3)}$$

Factor and simplify if possible.

EXAMPLE ⑥ PRACTICE PROBLEM

Simplify the following complex fractions.

a. $\dfrac{\dfrac{7}{x+8}+4}{\dfrac{6}{x-1}}$

b. $\dfrac{\dfrac{4}{x}+\dfrac{3}{y}}{\dfrac{2}{x^2}-\dfrac{6}{y^2}}$

7.4 Exercises

For Exercises 1 through 16, find the least common denominator for each set of rational expressions and write each expression in terms of the LCD.

1. $\dfrac{7}{12x^2y}$ $\dfrac{3}{28xy}$

2. $\dfrac{11a}{15b^2c}$ $\dfrac{14b}{25ac^3}$

3. $\dfrac{7m}{180n^2p}$ $\dfrac{4n}{150p}$

4. $\dfrac{16xz}{300y^4}$ $\dfrac{9}{84x^2y^2z}$

5. $\dfrac{x+2}{x+8}$ $\dfrac{x-4}{x+7}$

6. $\dfrac{x+9}{x-4}$ $\dfrac{x+7}{x+3}$

7. $\dfrac{x-5}{x+6}$ $\dfrac{x-5}{x+5}$

8. $\dfrac{x+1}{x-4}$ $\dfrac{x-3}{x-7}$

9. $2x$ $\dfrac{5}{x+1}$

10. $7y$ $\dfrac{x}{3y}$

11. $\dfrac{x+2}{(x+3)(x+7)}$ $\dfrac{x-4}{(x+3)(x+5)}$

12. $\dfrac{x-2}{(x+9)(x-3)}$ $\dfrac{x+7}{(x+4)(x-3)}$

13. $\dfrac{x+1}{(x-7)(x-2)}$ $\dfrac{x+2}{(x-8)(x-2)}$

14. $\dfrac{x+2}{x^2+5x+6}$ $\dfrac{x-4}{x^2-2x-8}$

15. $\dfrac{x+3}{x^2-2x-35}$ $\dfrac{x-7}{x^2+x-20}$

16. $\dfrac{x-5}{2x^2-9x-35}$ $\dfrac{x+6}{2x^2+11x+15}$

For Exercises 17 through 42, add or subtract the rational expressions.

17. $\dfrac{x+2}{x+7}+\dfrac{x+8}{x+7}$

18. $\dfrac{x-3}{x-9}+\dfrac{x-15}{x-9}$

19. $\dfrac{x+7}{x+5}-\dfrac{4x+22}{x+5}$

20. $\dfrac{x+7}{x-2}-\dfrac{x-6}{x-2}$

21. $\dfrac{x+5}{x+3}+\dfrac{x-2}{x+7}$

22. $\dfrac{x-4}{x+6}+\dfrac{x+3}{x-5}$

23. $\dfrac{x+2}{x+3}-\dfrac{x+4}{x+7}$

24. $\dfrac{x-7}{x+2}-\dfrac{x+3}{x+4}$

25. $5+\dfrac{2}{x+7}$

26. $8-\dfrac{3x}{x+4}$

27. $2x+\dfrac{7}{xy^2}$

28. $5y-\dfrac{6}{x^2}$

29. $\dfrac{x + 2}{(x + 5)(x + 3)} + \dfrac{x + 4}{(x + 5)(x + 3)}$

30. $\dfrac{3x - 6}{(2x + 7)(x + 3)} - \dfrac{7x + 8}{(2x + 7)(x + 3)}$

31. $\dfrac{x + 1}{(x - 7)(x - 2)} + \dfrac{x + 2}{(x - 8)(x - 2)}$

32. $\dfrac{2x + 5}{(x - 11)(x + 3)} + \dfrac{x - 3}{(x - 11)(x - 4)}$

33. $\dfrac{x - 2}{(x + 4)(x - 6)} - \dfrac{3x + 1}{(x + 3)(x - 6)}$

34. $\dfrac{x - 7}{(3x + 2)(2x + 5)} - \dfrac{x + 4}{(2x + 5)(x - 9)}$

35. $\dfrac{x + 3}{x^2 - 9x + 14} - \dfrac{x - 4}{x^2 - 10x + 16}$

36. $\dfrac{x + 2}{x^2 + 10x + 21} - \dfrac{x - 4}{x^2 + 8x + 15}$

37. $\dfrac{x - 2}{x^2 + 6x - 27} + \dfrac{x + 7}{x^2 + x - 12}$

38. $\dfrac{x - 5}{2x^2 - 9x - 35} + \dfrac{x + 6}{2x^2 + 11x + 15}$

39. $\dfrac{x + 2}{2x^2 + 7x + 3} + \dfrac{x - 4}{4x^2 + 7x - 15}$

40. $\dfrac{3x}{3x^2 + 24x + 36} + \dfrac{x^2}{2x^2 + 16x + 24}$

41. $\dfrac{5x + 43}{x^2 + 6x - 55} - \dfrac{2x + 10}{x^2 + 6x - 55}$

42. $\dfrac{x + 2}{6x^2 - 29x + 28} - \dfrac{x - 5}{2x^2 - 13x + 42}$

For Exercises 43 through 52, simplify the complex fractions.

43. $\dfrac{\dfrac{3}{x} + \dfrac{5}{y^2}}{\dfrac{7}{y} - \dfrac{6}{x^4}}$

44. $\dfrac{\dfrac{7}{x^3} - 4}{\dfrac{x}{y} + \dfrac{2}{x}}$

45. $\dfrac{\dfrac{2x}{5y^2} + \dfrac{6}{x}}{\dfrac{8}{x} + \dfrac{2}{3y}}$

46. $\dfrac{\dfrac{x}{y} - \dfrac{y}{x}}{\dfrac{x}{y^3} + \dfrac{y}{x^2}}$

47. $\dfrac{\dfrac{3}{x + 5}}{\dfrac{2}{x + 1} + 4x}$

48. $\dfrac{\dfrac{7}{x + 2}}{\dfrac{3}{x + 4} - 5x}$

49. $\dfrac{7 + \dfrac{2}{x + 8}}{\dfrac{1}{x - 3}}$

50. $\dfrac{10 - \dfrac{3}{2x + 5}}{\dfrac{7}{3x - 4}}$

51. $\dfrac{\dfrac{2}{x + 3} - \dfrac{5x}{x + 2}}{\dfrac{4}{x + 2} + \dfrac{7x}{x - 3}}$

52. $\dfrac{\dfrac{7}{x + 4} - \dfrac{6x}{x + 5}}{\dfrac{3}{x - 4} - \dfrac{2x}{x - 3}}$

53. In electronics if three resistors are connected in parallel, their combined resistance can be represented by

$$R = \dfrac{1}{\dfrac{1}{R_1} + \dfrac{1}{R_2} + \dfrac{1}{R_3}}$$

where R is the total resistance and R_1, R_2 and R_3 are the resistance of the three resistors.

a. Simplify the complex fraction on the right side of the equation.

b. If R_1 is 3 ohms, R_2 is 6 ohms, and R_3 is 9 ohms, then find R.

54. Use the simplified complex fraction you found in Exercise 53 part a to find R if f R_1 is 2 ohms, R_2 is 4 ohms, and R_3 is 8 ohms.

55. In photography the focal length inside a camera needed to focus an object in front of the lens is related to the distance between the object and the lens and the distance between the lens and the film in the camera. This relationship can be represented by

$$f = \dfrac{1}{\dfrac{1}{d} + \dfrac{1}{c}}$$

where f is the focal length, d is the distance between the object being photographed and the lens, and c is the distance between the lens and the film inside the camera.

a. Simplify the complex fraction on the right side of the equation.

b. If the distance from the lens to the object being photographed is 5 feet, and the distance from the lens to the film is 0.75 feet, find the focal length.

56. Use the simplified complex fraction you found in Exercise 55 part a to find the focal length if the distance from the lens to the object being photographed is 16 feet, and the distance from the lens to the film is 0.25 feet.

Solving Rational Equations

In this section we will learn to solve rational equations. Most often, you will want to eliminate all the fractions by multiplying both sides of the equation by the common denominator. This will allow you to solve the remaining equation using the techniques you learned in earlier chapters. When solving rational equations, you should keep in mind that some solutions will not be valid. When variables are in the denominator of a fraction, we must be very careful not to allow division by zero to take place. To avoid this, always check your solutions in the original equation to be sure that no invalid solutions are kept.

SOLVING RATIONAL EQUATIONS

1. Multiply both sides by the common denominator.

2. Solve the remaining equation.

3. Check the solution(s). (Watch for division by zero.)

EXAMPLE ☐ SOLVING AN AVERAGE COST EQUATION

The math club is planning a ski trip to a local mountain resort. The students want to keep the costs for each person as low as possible, so they plan to rent a van and split the costs evenly between them. They have decided that the person driving the van should not have to pay. If the van is going to cost $130 to rent and can hold up to 15 people, how many people need to go in the van to keep the transportation costs per person at $12?

© Agence Images/Alamy

Solution

Let p be the number of people going on the ski trip in the rented van, and let C be the cost per person going on the trip in dollars per person. Because the driver will not be paying for the van expenses, we need to subtract the driver from the total number of people p. If we take the total cost for the van and divide it by the number of people paying, we get the cost per person. This leads us to the following equation.

$$C(p) = \frac{130}{p - 1}$$

Because we want the average cost to be $12 per person, we can replace $C(p)$ with 12 and solve. To solve this equation, we need to get the variable p out of the denominator of the fraction. We can do this by multiplying both sides of the equation by the denominator $p - 1$.

$$\frac{130}{p - 1} = 12$$

$$(p - 1)\left(\frac{130}{p - 1}\right) = 12(p - 1) \quad \text{Multiply both sides by the denominator.}$$

$$130 = 12p - 12 \quad \text{Solve for } p.$$

$$142 = 12p$$

$$\frac{142}{12} = p$$

$$11.83 \approx p$$

Therefore the math club needs to have at least 12 people go in the van to keep the transportation cost at $12 per person or less. Note that we had to round this answer up, or the cost would have been more than $12 per person. In this type of situation you would always want to round up to the next whole person.

EXAMPLE ① PRACTICE PROBLEM

In Section 7.1 we found the model $a(m) = \dfrac{50}{m}$ for the acceleration a in m/s^2 that a fixed amount of force would give when applied to a mass of m kg.

a. If an object has an acceleration of 4 m/s^2 when this force is applied, what is the object's mass?

b. What mass can be accelerated at 0.5 m/s^2 using the given fixed amount of force?

For rational functions that arise from dividing two functions the technique of multiplying both sides of the equation by the denominator quickly clears the fraction and allows you to solve the remaining equation. Although some of these equations might look very messy, the use of a graphing calculator can help to manage the solution process.

EXAMPLE 2 FINDING THE AVERAGE DISBURSEMENTS

In Section 7.1 we found a model for the average disbursements made by the U.S. Medicare program.

$$P(t) = \frac{-3160.64t^2 + 52536.5t - 4548.31}{0.39t + 35.65}$$

where $P(t)$ represents the average dollars disbursed per person by Medicare t years since 1990. In what year did the average dollars disbursed hit $5500?

Solution

We are given an average disbursement, so we can substitute $P(t) = 5500$.

$$5500 = \frac{-3160.64t^2 + 52536.5t - 4548.31}{0.39t + 35.65}$$

$$(0.39t + 35.65)(5500) = \frac{-3160.64t^2 + 52536.5t - 4548.31}{0.39t + 35.65} \cdot (0.39t + 35.65)$$

$$2145t + 196075 = -3160.64t^2 + 52536.5t - 4548.31$$

$$0 = -3160.64t^2 + 50391.5t - 200623.31$$

$$\approx 7.7 \qquad t \approx 8.24 \qquad \textbf{By the quadratic formula.}$$

Because both $t = 7.7$ and $t = 8.24$ represent the year 1998, we have that in 1998 the average disbursement per person made by Medicare was about $5500. We can check these solutions using the graph or table.

Once you have multiplied by the denominator, you can combine all the like terms. In the case of Example 2 you were left with a quadratic equation and therefore can use the quadratic formula or factoring to solve. In general, multiplying by the common denominator will allow you to solve most equations involving rational expressions.

EXAMPLE   **3**   SOLVING SIMPLE RATIONAL EQUATIONS

Solve the following rational equations.

a. $\dfrac{3}{x + 2} = 12$

b. $\dfrac{10}{x - 8} = 5$

c. $\dfrac{5}{x + 1} = 3 - \dfrac{7}{x + 1}$

Solution

a.
$$\frac{3}{x + 2} = 12$$

$$(x + 2)\left(\frac{3}{x + 2}\right) = 12(x + 2) \qquad \textbf{Multiply both sides of the equation by the denominator.}$$

$$3 = 12x + 24 \qquad \textbf{Solve for } x.$$

$$-21 = 12x$$

$$-\frac{21}{12} = x$$

Check the solution using a graph and trace.

$$-\frac{7}{4} = x$$

b.
$$\frac{10}{x - 8} = 5$$

$(x - 8)\left(\frac{10}{x - 8}\right) = 5(x - 8)$ Multiply both sides of the equation by the denominator.

$10 = 5x - 40$ Solve for x.

$50 = 5x$

$10 = x$

X	Y1	Y2
10	5	5

Y1■10/(X-8)

Check the solution using the table.

c.
$$\frac{5}{x + 1} = 3 - \frac{7}{x + 1}$$

$(x + 1)\left(\frac{5}{x + 1}\right) = \left(3 - \frac{7}{x + 1}\right)(x + 1)$ Multiply both sides by the common denominator.

$5 = 3(x + 1) - \left(\frac{7}{x + 1}\right)(x + 1)$ Use the distributive property.

$5 = 3x + 3 - 7$ Solve for x.

$5 = 3x - 4$

$9 = 3x$

$x = 3$

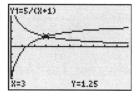

Y1=5/(X+1)

X=3   Y=1.25

Check the solution.

EXAMPLE ③ PRACTICE PROBLEM

Solve the following rational equations.

a. $\dfrac{7}{x - 4} = 14$

b. $\dfrac{35}{x + 3} = 4 + \dfrac{15}{x + 3}$

This solving technique can be used with more complicated rational equations. When there is more than one denominator, it is quickest if you first find a common denominator for all fractions in the equation. When you multiply by the common denominator, you will eliminate all of the fractions, and you should then be able to solve the remaining equation. Remember to check your solutions for values that might not be valid.

EXAMPLE 4 SOLVING RATIONAL EQUATIONS

Solve the following rational equations and check your solution(s).

a. $\dfrac{12}{x + 3} = \dfrac{7}{x - 5}$

b. $\dfrac{4}{x - 4} = \dfrac{3}{x + 7}$

c. $\dfrac{7}{x + 5} + \dfrac{3x}{x - 2} = \dfrac{42}{x^2 + 3x - 10}$

d. $\dfrac{2}{x + 3} + \dfrac{7}{x + 4} = \dfrac{4x - 11}{x^2 + 7x + 12}$

Solution

a.
$$\frac{12}{x + 3} = \frac{7}{x - 5}$$

$$(x + 3)(x - 5)\left(\frac{12}{x + 3}\right) = \left(\frac{7}{x - 5}\right)(x + 3)(x - 5)$$ Multiply both sides by the common denominator.

$$(x - 5)12 = 7(x + 3)$$ Cancel denominators and

$$12x - 60 = 7x + 21$$ distribute the multiplication.

$$5x = 81$$ Solve for x.

$$x = \frac{81}{5}$$

$$x = 16.2$$

$$\frac{12}{(16.2) + 3} = \frac{7}{(16.2) - 5}$$ Check the solution.

$$0.625 = 0.625$$

b.
$$\frac{4}{x - 4} = \frac{3}{x + 7}$$

$$(x - 4)(x + 7)\left(\frac{4}{x - 4}\right) = \left(\frac{3}{x + 7}\right)(x - 4)(x + 7)$$ Multiply both sides by the common denominator.

$$(x + 7)(4) = (3)(x - 4)$$

$$4x + 28 = 3x - 12$$ Solve for x.

$$x = -40$$

$$\frac{4}{(-40) - 4} = \frac{3}{(-40) + 7}$$ Check the solution.

$$-0.091 = -0.091$$

c.

$$\frac{7}{x+5} + \frac{3x}{x-2} = \frac{42}{x^2+3x-10}$$

$$\frac{7}{x+5} + \frac{3x}{x-2} = \frac{42}{(x+5)(x-2)}$$

$$(x+5)(x-2)\left(\frac{7}{x+5} + \frac{3x}{x-2}\right) = \left(\frac{42}{(x+5)(x-2)}\right) \cdot (x+5)(x-2)$$

Note that you cannot cancel on the left side until you distribute.

$$(x+5)(x-2)\left(\frac{7}{x+5}\right) + (x+5)(x-2)\left(\frac{3x}{x-2}\right) = 42$$

$$7(x-2) + 3x(x+5) = 42$$

$$7x - 14 + 3x^2 + 15x = 42 \qquad \text{Solve for } x.$$

$$3x^2 + 22x - 56 = 0$$

$$x = 2 \qquad x = -\frac{28}{3}$$

X	Y₁	Y₂
2	ERROR	ERROR
-9.333	.8552	.8552

X=

Check the solutions.
Because 2 results in division by zero, it is not a valid solution.

Because $x = 2$ causes some of the fractions to have division by zero, we must eliminate that solution; thus we have only $x = -\dfrac{28}{3}$ as a solution.

d.

$$\frac{2}{x+3} + \frac{7}{x+4} = \frac{4x-11}{x^2+7x+12} \qquad \begin{array}{l}\text{Factor to find the}\\\text{common denominator.}\end{array}$$

$$\frac{2}{x+3} + \frac{7}{x+4} = \frac{4x-11}{(x+3)(x+4)}$$

$$(x+3)(x+4)\left(\frac{2}{x+3} + \frac{7}{x+4}\right) = \left(\frac{4x-11}{(x+3)(x+4)}\right)(x+3)(x+4)$$

$$(x+3)(x+4)\left(\frac{2}{x+3}\right) + (x+3)(x+4)\left(\frac{7}{x+4}\right) = 4x-11$$

$$(x+4)(2) + (x+3)(7) = 4x - 11$$

$$2x + 8 + 7x + 21 = 4x - 11$$

$$9x + 29 = 4x - 11$$

$$5x = -40$$

$$x = -8 \qquad \text{Check the solution.}$$

$$\frac{2}{(-8)+3} + \frac{7}{(-8)+4} = \frac{4(-8)-11}{(-8)^2 + 7(-8) + 12}$$

$$-2.15 = -2.15$$

EXAMPLE ④ PRACTICE PROBLEM

Solve the following rational equations.

a. $\dfrac{3}{x + 2} = \dfrac{6}{x + 9}$

b. $\dfrac{5x}{x - 4} + \dfrac{3}{x + 2} = \dfrac{-18}{x^2 - 2x - 8}$

c. $\dfrac{5}{x + 2} = 8 - \dfrac{52}{x + 7}$

7.5 Exercises

1. The state of California's spending per person has increased dramatically since 1950. The state's population from 1950 can be modeled by

   $$P(t) = 0.464t - 12.47$$

   where $P(t)$ is California's population in millions of people t years since 1900. The amount that California spent in millions of dollars can be modeled by

   $$S(t) = 55.125t^2 - 6435.607t + 186914.286$$

   where $S(t)$ is the amount California spent in millions of dollars t years since 1900.

   Source: Models derived from data in the Governor's Budget Summary as printed in the North County Times *Feb. 9, 2003.*

   a. Find a new function that gives the spending per capita (per person) t years since 1900.

   b. Find the per capita spending in California in 1995.

   c. Find the year in which the per capita spending in California reached $3000 per person.

2. The Fancy Affair catering company is catering an event for a local charity. Jan, the owner of Fancy Affair, is donating her time to the cause but needs to charge the charity for the food, decorations, and supplies used. If p people attend the charity event, Jan has figured her total cost to be modeled by

   $$C(p) = 4.55p + 365.00$$

   where C is the total cost in dollars for the food, decorations, and supplies when p people attend the charity event.

   a. Find a new function that gives the per person cost for p people to attend the charity event.

   b. Use your new model to find how many people need to attend to keep the per person costs down to $7.50 each.

3. The percent of housing units in Colorado that were vacant in the 1990s can be modeled by

   $$P(t) = \dfrac{216.89t^2 - 2129.2t + 19114.9}{33.67t + 1444.29}$$

   where $P(t)$ is the percent of housing units in Colorado that were vacant t years since 1990.

   Source: Model derived from data from the U.S. Department of Commerce Bureau of Economic Analysis.

   a. Find the percent of housing units vacant in 1995.

   b. Find when there was an 11.5% vacancy rate for housing units in Colorado.

4. The per capita personal spending in the United States can be modeled by

   $$s(t) = \dfrac{248158.9t + 186907.6}{0.0226t^2 + 1.885t + 204.72}$$

   where $s(t)$ represents the per capita personal spending in dollars in the United States t years since 1970.

Source: Model derived from data from the U.S. Department of Commerce Bureau of Economic Analysis.

   a. Find the per capita personal spending in the United States in 1995.

   b. In what year did the per capita personal spending in the United States reach $25,000?

5. The population of the United States since 2000 can be modeled by

$$P(t) = 2.74t + 282.63$$

where $P(t)$ represents the population of the United States in millions t years since 2000. The national debt of the United States in millions of dollars

t years since 2000 can be modeled by

$$D(t) = 540,872.6t + 5,109,943.3$$

where $D(t)$ represents the national debt of the U.S. in millions of dollars t years since 2000.

$ 9,524,810,769,380.27

As of July 21, 2008, 8:31:30 PM GMT

Source: Models derived from data from the U.S. Department of Commerce Bureau of Economic Analysis.

   a. Find a new model for the average amount of national debt per person in the United States t years since 2000.

   b. Find the average amount of national debt per person in 2015.

   c. Find when the average amount of national debt per person will reach $40,000 per person.

6. The average amount of benefits received by people in the U.S. food stamp program can be modeled by

$$B(t) = \frac{-470001t^2 + 4110992t + 14032612}{-469.4t^2 + 3745t + 19774}$$

where $B(t)$ is the average benefit in dollars per person for people in the U.S. food stamp program t years since 1990.

Source: Model derived from data from Statistical Abstract 2001.

   Find when the average benefit was $800 per person.

7. The weight of a body varies inversely as the square of its distance from the center of the earth. If the radius

of the earth is 4000 miles, how far from the earth's surface must a 220-lb man be to weigh only 80 lb?

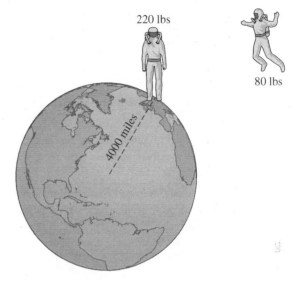

220 lbs

80 lbs

4000 miles

8. In electronics, Ohm's Law states that if the voltage in a circuit is constant, the current that flows through the circuit is inversely proportional to the resistance of that circuit. When the resistance R is 200 ohms, the current I is 1.2 amp. Find the resistance when the current is 0.9 amp. (*Note:* The proportionality constant in this problem is actually the voltage supplied to the circuit.)

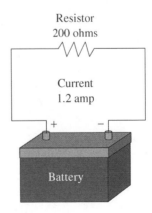

Resistor
200 ohms

Current
1.2 amp

+ −

Battery

9. The illumination of a light source is inversely proportional to the square of the distance from the light source. If a light has an illumination of 25.5 foot-candles at a distance of 10 feet from the light source:

   a. Find a model for the illumination of this light.

b. What is the illumination of this light at a distance of 5 feet from the light source?

c. How far away from this light source would you need to be for the illumination to be 50 foot-candles?

10. a. If y varies inversely with the cube of x and $y = 11$ when $x = 5$, find an equation to represent this relationship.

b. Find x if $y = 4$.

11. a. If m varies inversely with n^5 and $m = 25$ when $n = 2$, find an equation to represent this relationship.

b. Find m if $n = 4$.

c. Find n if $m = 50$.

12. a. If W varies inversely with p^4 and $W = 135$ when $p = 3$, find an equation to represent this relationship.

b. Find W when $p = 10$.

c. Find p when $W = 400$.

For Exercises 13 through 42, solve and check with a table or graph.

13. $2 - \dfrac{6}{x} = \dfrac{4}{x}$

14. $3 + \dfrac{5}{2x} = \dfrac{4}{x}$

15. $5 = \dfrac{4}{2w + 3}$

16. $10 = \dfrac{6}{3p + 4}$

17. $2t = \dfrac{8}{t - 3}$

18. $5x = \dfrac{-10}{3x - 7}$

19. $3m = \dfrac{-9}{m + 4}$

20. $2x = \dfrac{-7}{x + 6}$

21. $\dfrac{x + 5}{3x - 6} = \dfrac{2}{3}$

22. $\dfrac{2t - 3}{5t + 6} = \dfrac{11}{17}$

23. $\dfrac{x^2}{x + 5} = \dfrac{25}{x + 5}$

24. $\dfrac{x^2}{x + 6} = \dfrac{36}{x + 6}$

25. $\dfrac{w^2}{w - 3} = \dfrac{9}{w - 3}$

26. $\dfrac{x^2}{x - 7} = \dfrac{49}{x - 7}$

27. $\dfrac{12}{x + 5} - 2 = \dfrac{-10}{x + 3}$

28. $\dfrac{8}{k + 3} + 5 = \dfrac{-21}{k - 4}$

29. $\dfrac{v + 6}{v - 2} = \dfrac{v - 1}{v - 8}$

30. $\dfrac{x + 9}{x - 3} = \dfrac{x + 4}{x - 6}$

31. $\dfrac{d + 3}{d + 5} = \dfrac{d + 8}{d + 2}$

32. $\dfrac{n - 8}{n - 5} = \dfrac{n + 9}{n - 2}$

33. $\dfrac{x}{x - 14} + \dfrac{9}{x - 7} = \dfrac{x^2}{x^2 - 21x + 98}$

34. $\dfrac{p}{p - 8} + \dfrac{6}{p - 4} = \dfrac{p^2}{p^2 - 12p + 32}$

35. $\dfrac{k - 2}{k + 1} = \dfrac{7k + 22}{k^2 + 6k + 5}$

36. $\dfrac{m + 4}{m + 1} = \dfrac{30}{m^2 - 2m - 3}$

37. $\dfrac{2x + 5}{x - 7} = \dfrac{x - 2}{x + 3}$

38. $\dfrac{3c - 7}{c + 5} = \dfrac{2c + 3}{c - 8}$

39. $\dfrac{t}{t - 6} + \dfrac{5}{t - 3} = \dfrac{t^2}{t^2 - 9t + 18}$

40. $\dfrac{x}{x - 4} + \dfrac{4}{x - 2} = \dfrac{x^2}{x^2 - 6x + 8}$

41. $\dfrac{25}{w^2 + w - 12} = \dfrac{w + 4}{w - 3}$

42. $\dfrac{c - 3}{c + 4} = \dfrac{14}{c^2 + 6c + 8}$

Chapter 7 Summary

Section 7.1 Introduction to Rational Functions

- A rational expression is anything of the form

$$\frac{P(x)}{Q(x)}$$

  where $P(x)$ and $Q(x)$ are polynomials and $Q(x) \neq 0$.

- Many rational functions come from the division necessary to calculate averages.
- The domain of a rational function that is not in context is found by setting the denominator equal to zero.
- The domain will be all real numbers except where the denominator equals zero.
- The variable y varies directly with x if

$$y = kx$$

- The variable y varies inversely with x if

$$y = \frac{k}{x}$$

- For direct and inverse variation k is called the variation constant and can be any nonzero real number.

EXAMPLE

a. If y varies inversely with x^5 and $y = 7$ when $x = 4$, find an equation to represent this relationship.

b. Find y if $x = 6$.

Solution

a. $$y = \frac{k}{x^5}$$

$$7 = \frac{k}{4^5}$$ **Substitute the values for x and y and find k.**

$$7 = \frac{k}{1024}$$

$$7168 = k$$

$$y = \frac{7168}{x^5}$$ **Write the equation.**

b.

$$y = \frac{7168}{x^5}$$

$$y = \frac{7168}{6^5}$$

$$y = 0.922$$

EXAMPLE 2

Give the domain of the rational function

$$f(x) = \frac{5x + 7}{(x + 3)(x - 7)}$$

Solution

$$(x + 3)(x - 7) = 0$$

$$x + 3 = 0 \qquad x - 7 = 0$$ Set the denominator equal to zero and solve.

$$x = -3 \qquad x = 7$$

The domain of $f(x)$ is all real numbers except $x \neq -3$ or 7.

Section 7.2 Simplifying Rational Expressions

- To simplify rational expressions, factor the numerator(s) and denominator(s) so that any common factors can be canceled out.

- Always use caution in canceling in rational expressions. You can cancel only if the entire numerator and denominator is factored (written as a multiplication). If there are any separate terms, you cannot cancel.

- Long division of polynomials is done in the same way you divide large numbers using long division.

- If terms are missing in a divisor or dividend replace them with a zero term so that the process will be less confusing.

- Synthetic division can be used to simplify long division problems when the divisor can be written in the form $x - c$.

EXAMPLE 3

Simplify the rational expression $\dfrac{5x + 20}{x^2 + 5x + 4}$.

Solution

$$\frac{5x + 20}{x^2 + 5x + 4} = \frac{5(x + 4)}{(x + 1)(x + 4)}$$ Factor the denominator to find common factors that can be canceled.

$$= \frac{5\cancel{(x + 4)}}{(x + 1)\cancel{(x + 4)}}$$

$$= \frac{5}{x + 1}$$

EXAMPLE 4

Use long division to divide $(3x^3 - 12x + 144) \div (x + 4)$.

Solution

$$
\begin{array}{r}
3x^2 - 12x + 36 \\
x + 4 \overline{)3x^3 + 0x^2 - 12x + 144} \\
\underline{3x^3 + 12x^2} \\
-12x^2 - 12x \\
\underline{-12x^2 - 48x} \\
36x + 144 \\
\underline{36x + 144} \\
0
\end{array}
$$

Include a zero term for the missing squared term in the dividend.

Watch the signs when you subtract.

There is not remainder.

Therefore $(3x^3 - 12x + 144) \div (x + 4) = 3x^2 - 12x + 36$

Section 7.3 Multiplying and Dividing Rational Expressions

- When multiplying or dividing rational expressions, factor the numerator(s) and denominator(s) so that any common factors can be canceled out as the first step.
- Multiply rational expressions in the same way you multiply fractions. Multiply the numerators together and multiply the denominators together. Then simplify further if possible.
- Divide rational expressions in the same way you divide fractions. Multiply by the reciprocal of the divisor.
- The result from the multiplication or division can be left in factored form.

EXAMPLE 5

Perform the indicated operation and simplify.

a. $\dfrac{x^2 + 7x + 10}{x + 3} \cdot \dfrac{x - 7}{x^2 + x - 20}$

b. $\dfrac{x^2 + 2x - 15}{2x^2 + 13x + 15} \div \dfrac{x^2 + 4x - 21}{2x^2 + 23x + 63}$

Solution

a. $\dfrac{x^2 + 7x + 10}{x + 3} \cdot \dfrac{x - 7}{x^2 + x - 20}$

Factor the numerators and denominators.

$\dfrac{(x + 5)(x + 2)}{x + 3} \cdot \dfrac{x - 7}{(x + 5)(x - 4)}$

Cancel out common factors.

$\dfrac{(\cancel{x + 5})(x + 2)}{x + 3} \cdot \dfrac{x - 7}{(\cancel{x + 5})(x - 4)}$

$\dfrac{(x + 2)(x - 7)}{(x + 3)(x - 4)}$

b. $\dfrac{x^2 + 2x - 15}{2x^2 + 13x + 15} \div \dfrac{x^2 + 4x - 21}{2x^2 + 23x + 63}$

$\dfrac{x^2 + 2x - 15}{2x^2 + 13x + 15} \cdot \dfrac{2x^2 + 23x + 63}{x^2 + 4x - 21}$ Multiply by the reciprocal of the divisor.

$\dfrac{(x + 5)(x - 3)}{(2x + 3)(x + 5)} \cdot \dfrac{(2x + 9)(x + 7)}{(x - 3)(x + 7)}$ Factor the numerators and denominators.

$\dfrac{\cancel{(x + 5)}\cancel{(x - 3)}}{(2x + 3)\cancel{(x + 5)}} \cdot \dfrac{(2x + 9)\cancel{(x + 7)}}{\cancel{(x - 3)}\cancel{(x + 7)}}$ Cancel out the common factors.

$\dfrac{2x + 9}{2x + 3}$ Note that the $2x$'s do not cancel because they are terms and not factors.

Section 7.4 Adding and Subtracting Rational Expressions

- To add or subtract rational expressions, you need a common denominator.
- The least common denominator (LCD) can be found by factoring all the denominators and taking the highest power of each factor for the LCD.
- Once a common denominator is found, write each fraction in terms of the common denominator by multiplying both the numerator and the denominator by any missing factors of the common denominator.
- To add or subtract rational expressions that have a common denominator, add or subtract the numerators and keep the denominator the same.
- Simplify complex fractions using the LCD of all the fractions in the complex rational expression.

EXAMPLE 6

Perform the indicated operation and simplify.

a. $\dfrac{x + 2}{x^2 - 3x - 28} + \dfrac{x - 3}{x^2 - 9x + 14}$

b. $\dfrac{x + 7}{x^2 + 8x + 15} - \dfrac{x + 4}{2x^2 + 6x - 20}$

Solution

a. $\dfrac{x + 2}{x^2 - 3x - 28} + \dfrac{x - 3}{x^2 - 9x + 14}$

$\dfrac{x + 2}{(x + 4)(x - 7)} + \dfrac{x - 3}{(x - 2)(x - 7)}$ Factor the denominators to find the LCD.

$\dfrac{x - 2}{x - 2} \cdot \dfrac{x + 2}{(x + 4)(x - 7)} + \dfrac{x - 3}{(x - 2)(x - 7)} \cdot \dfrac{x + 4}{x + 4}$ Multiply by the missing factors.

$\dfrac{x^2 - 4}{(x - 2)(x + 4)(x - 7)} + \dfrac{x^2 + x - 12}{(x - 2)(x + 4)(x - 7)}$

$\dfrac{2x^2 + x - 16}{(x - 2)(x + 4)(x - 7)}$ Add the numerators.

b.
$$\frac{x + 7}{x^2 + 8x + 15} - \frac{x + 4}{2x^2 + 6x - 20}$$

Factor the denominators to find the LCD.

$$\frac{x + 7}{(x + 3)(x + 5)} - \frac{x + 4}{(x - 2)(x + 5)}$$

Multiply by the missing factors.

$$\frac{x - 2}{x - 2} \cdot \frac{x + 7}{(x + 3)(x + 5)} - \frac{x + 4}{(x - 2)(x + 5)} \cdot \frac{x + 3}{x + 3}$$

$$\frac{x^2 + 5x - 14}{(x - 2)(x + 3)(x + 5)} - \frac{x^2 + 7x + 12}{(x - 2)(x + 3)(x + 5)}$$

Subtract the numerators.

$$\frac{-2x - 26}{(x - 2)(x + 3)(x + 5)}$$

$$\frac{-2(x + 13)}{(x - 2)(x + 3)(x + 5)}$$

Section 7.5 Solving Rational Equations

- To solve an equation that contains rational expressions, you can follow these steps:

  1. Multiply both sides of the equation by the least common denominator.
  2. Solve the remaining equation.
  3. Check your solution(s).

- Keep in mind when solving rational equations that some potential solutions will not work. Always check for division by zero in the original equation.

EXAMPLE 7

Solve the following equations.

a. $\dfrac{3}{x + 5} = 4 - \dfrac{7}{x + 5}$

b. $\dfrac{2x}{x + 3} + \dfrac{5}{x - 7} = \dfrac{8x - 6}{x^2 - 4x - 21}$

Solution

a.
$$\frac{3}{x + 5} = 4 - \frac{7}{x + 5}$$

$$(x + 5) \cdot \frac{3}{x + 5} = \left(4 - \frac{7}{x + 5}\right) \cdot (x + 5)$$

$$(x + 5) \cdot \frac{3}{x + 5} = 4(x + 5) - \frac{7}{x + 5} \cdot (x + 5)$$

Multiply by the common denominator.

$$(\cancel{x + 5}) \cdot \frac{3}{\cancel{x + 5}} = 4(x + 5) - \frac{7}{\cancel{x + 5}} \cdot (\cancel{x + 5})$$

$$3 = 4x + 20 - 7$$ Solve the remaining equation.

$$3 = 4x + 13$$

$$-10 = 4x$$

$$-2.5 = x$$

X	Y₁	Y₂
-2.5	1.2	1.2

X=

Check your answer.

b. $\dfrac{2x}{x+3} + \dfrac{5}{x-7} = \dfrac{8x-6}{x^2 - 4x - 21}$

$\dfrac{2x}{x+3} + \dfrac{5}{x-7} = \dfrac{8x-6}{(x+3)(x-7)}$ Factor to find the LCD.

$(x+3)(x-7) \cdot \left(\dfrac{2x}{x+3} + \dfrac{5}{x-7} \right)$ Multiply both sides by the LCD.

$= \dfrac{8x-6}{(x+3)(x-7)} \cdot (x+3)(x-7)$

$(x+3)(x-7) \cdot \dfrac{2x}{x+3} + (x+3)(x-7) \cdot \dfrac{5}{x-7}$

$= \dfrac{8x-6}{(x+3)(x-7)} \cdot (x+3)(x-7)$

$(x+3)(x-7) \cdot \dfrac{2x}{x+3} + (x+3)(x-7) \cdot \dfrac{5}{x-7}$

$= \dfrac{8x-6}{(x+3)(x-7)} \cdot (x+3)(x-7)$

$2x(x-7) + 5(x+3) = 8x - 6$ Solve the remaining equation.

$2x^2 - 14x + 5x + 15 = 8x - 6$

$2x^2 - 17x + 21 = 0$

$(2x - 3)(x - 7) = 0$

$2x - 3 = 0 \qquad x - 7 = 0$ The 7 causes division by zero.

$x = \dfrac{3}{2} \qquad x = 7$ Only one answer works.

$x = 1.5$ Check your answers.

X	Y₁	Y₂
1.5	-.2424	-.2424
7	ERROR	ERROR

X=

Chapter 7 Review Exercises

1. The illumination of a light source is inversely proportional to the square of the distance from the light source. A certain light has an illumination of 40 foot-candles at a distance of 30 feet from the light source.

    a. Find a model for the illumination of this light.

    b. What is the illumination of this light at a distance of 40 feet from the light source?

    c. At what distance would the illumination of this light be 50 foot-candles? **[7.1]**

2. a. If y varies inversely with x^5 and $y = 7$ when $x = 2$, find an equation to represent this relationship.

    b. Find y if $x = 4$. **[7.1]**

3. Math R Us (MRU) is having a math competition and wants to have a total of $500 in prize money. To run the competition, MRU will need to have the prize money plus $7.50 per person competing in the competition. They want to set an entrance fee for each competitor that will cover their costs. MRU plans to give 10 scholarships to those who cannot afford to pay.

    a. Find a model that gives the entrance fee MRU should charge if they have p people compete. (p includes the 10 people who are on scholarship and therefore will not pay an entrance fee.) **[7.1]**

    b. Use your model to estimate the entrance fee MRU should charge if 100 people compete. **[7.1]**

    c. If MRU wants to have an entrance fee of $10, how many people do they need to participate in this competition? **[7.5]**

4. The per capita amount of cheese consumed by Americans can be modeled by

$$C(t) = \frac{637.325t + 7428.649}{0.226t^2 + 10.925t + 332.82}$$

where $C(t)$ represents the per capita amount of cheese consumed by Americans in pounds per person t years since 1990.

Source: Models derived from data in Statistical Abstract 2001.

    a. Estimate the per capita amount of cheese consumed by Americans in 1995.

    b. Estimate when the per capita amount of cheese consumed by Americans will reach 30 pounds per person. **[7.5]**

For Exercises 5 through 10, find the domain of the given rational function.

5. $f(x) = \dfrac{3x + 2}{x - 9}$ **[7.1]**

6. $g(x) = \dfrac{x + 5}{(x + 5)(x - 3)}$ **[7.1]**

7. $h(x) = \dfrac{5x + 2}{2x^2 - 7x - 15}$ **[7.1]**

8. $f(x) = \dfrac{x + 3}{7x^2 + 5x + 25}$ **[7.1]**

9.

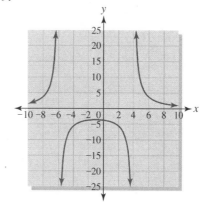

[7.1]

10.

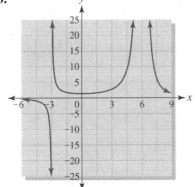

[7.1]

11. Simplify $\dfrac{4x + 8}{x^2 + 6x + 8}$ [7.2]

12. Simplify $\dfrac{6x^2 + 25x + 14}{2x^2 + 25x + 63}$ [7.2]

For Exercises 13 through 14, use long division to divide.

13. $(x^4 + 3x^3 + x^2 + 15x - 20) \div (x^2 + 5)$ [7.2]

14. $(3x^3 + 20x^2 - 3x - 10) \div (3x + 2)$ [7.2]

For Exercises 15 through 16, use synthetic division to divide.

15. $(4x^3 + 15x^2 + 12x - 4) \div (x + 2)$ [7.2]

16. $(5x^3 - 20x - 75) \div (x - 3)$ [7.2]

For Exercises 17 through 26, perform the indicated operation and simplify the rational expressions:

17. $\dfrac{x + 3}{x - 15} \cdot \dfrac{x - 15}{x - 2}$ [7.3]

18. $\dfrac{4}{x + 3} \cdot \dfrac{x + 9}{x - 7}$ [7.3]

19. $\dfrac{x + 2}{x - 3} \div \dfrac{x + 7}{x - 3}$ [7.3]

20. $\dfrac{(x + 4)(x + 5)}{(x - 5)(x + 2)} \div \dfrac{(x + 5)(x - 7)}{(x + 2)(x - 7)}$ [7.3]

21. $\dfrac{2n + 6}{n^2 + 8n + 15} \cdot \dfrac{3n + 15}{n^2 + 11x + 28}$ [7.3]

22. $\dfrac{v + 5}{v^2 - 6v - 55} \div \dfrac{v - 8}{v^2 - 4v - 77}$ [7.3]

23. $\dfrac{12}{x + 6} + \dfrac{4x + 3}{x + 6}$ [7.4]

24. $\dfrac{3}{x - 5} - \dfrac{7}{x + 2}$ [7.4]

25. $\dfrac{x + 2}{(x + 3)(x + 5)} + \dfrac{6}{(x + 3)(x - 7)}$ [7.4]

26. $\dfrac{2x - 7}{x^2 - 5x - 14} + \dfrac{x + 5}{x^2 - 4x - 21}$ [7.4]

For Exercises 27 through 34, solve the rational equations. Check your answers with a table or graph.

27. $\dfrac{5}{x - 3} = \dfrac{2}{x + 7}$ [7.5]

28. $\dfrac{4x}{x + 3} = \dfrac{2x}{x - 9}$ [7.5]

29. $\dfrac{2}{x + 5} + 6 = \dfrac{-5}{x - 3}$ [7.5]

30. $\dfrac{x}{x + 5} + \dfrac{7}{x - 3} = \dfrac{56}{(x + 5)(x - 3)}$ [7.5]

31. $\dfrac{w}{w - 7} - \dfrac{3}{w - 4} = \dfrac{9}{(w - 7)(w - 4)}$ [7.5]

32. $\dfrac{5x + 3}{x^2 + 7x - 9} = \dfrac{10x - 12}{x^2 + 7x - 9}$ [7.5]

33. $\dfrac{3k}{k + 2} + \dfrac{2}{k + 5} = \dfrac{50}{k^2 + 7k + 10}$ [7.5]

34. $\dfrac{5}{x^2 + 3x - 28} = \dfrac{3x}{x^2 - 8x + 16}$ [7.5]

Chapter 7 Test

1. The illumination of a light source is inversely proportional to the square of the distance from the light source. A certain light has an illumination of 30 foot-candles at a distance of 12 feet from the light source.

   a. Find a model for the illumination of this light.

   b. What is the illumination of this light at a distance of 7 feet from the light source?

   c. At what distance would the illumination of this light be 20 foot-candles?

b.

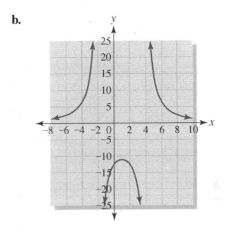

For Exercises 2 through 8, perform the indicated operation and simplify the rational expressions:

2. $\dfrac{2x}{x + 7} \cdot \dfrac{x + 3}{x - 4}$

3. $\dfrac{(x + 3)(x + 5)}{(x - 7)(x + 3)} \div \dfrac{(x + 5)(x + 2)}{(x + 4)(x - 7)}$

4. $\dfrac{5w + 3}{w^2 - 4w - 21} \cdot \dfrac{2w - 14}{5w^2 - 17w - 12}$

5. $\dfrac{2m + 4}{m^2 - m - 20} \div \dfrac{m + 2}{m^2 - 4m - 5}$

6. $\dfrac{5x}{2x - 7} + \dfrac{3x - 8}{2x - 7}$

7. $\dfrac{x + 5}{x - 3} - \dfrac{x - 2}{x + 4}$

8. $\dfrac{5x + 2}{x^2 + 5x + 6} + \dfrac{x - 4}{x^2 + 9x + 14}$

9. Find the domain for the rational functions.

   a. $h(x) = \dfrac{3x - 7}{x^2 + 6x - 27}$

10. Divide $(5x^3 + 22x^2 - 17x - 10) \div (5x + 2)$

11. Divide $(x^4 + 5x^3 + 3x^2 + 19x + 20) \div (x + 5)$

For Exercises 12 through 16, solve the rational equations: Check your answers with a table or graph.

12. $\dfrac{2x + 5}{x - 7} = \dfrac{3x - 12}{x - 7}$

13. $\dfrac{5}{x + 6} = \dfrac{3}{x - 4}$

14. $\dfrac{5}{x + 2} + 7 = \dfrac{-16}{x - 5}$

15. $\dfrac{x^2}{(x + 2)(x + 3)} + \dfrac{5}{x + 3} = \dfrac{4}{x + 2}$

16. $\dfrac{5.6}{x^2 - 3x - 10} = \dfrac{2x}{x^2 + 9x + 14}$

17. The total labor force in the state of Florida can be modeled by

$$L(t) = 141.5t + 7826.8$$

where $L(t)$ represents the labor force in thousands in the state of Florida t years since 2000. The number of people unemployed in Florida can be modeled by

$$E(t) = 80.5t + 287.5$$

where $E(t)$ represents the number of people in the workforce who are unemployed in thousands in the state of Florida. The unemployment rate is the percentage of people in the labor force who are not employed.

Source: Models derived from data obtained from the Florida Research and Economic Database 2003.

a. Find a model for the unemployment rate for the state of Florida.

b. Estimate the unemployment rate in Florida during 2003.

c. Estimate when the unemployment rate in Florida will reach 8%.

Chapter 7 Projects

HOW BRIGHT IS THAT LIGHT?

In this project you will investigate the inverse relationship between the illumination of a light and the square of the distance from the light source. Fix the position of the flashlight and place the tape measure out from the front of the flashlight. You will want to collect eight or more data values for the light's illumination using the CBL light sensor probe at different distances. Measure your distances in feet, but remember that you can use fractions of feet in your measurements. Be sure to turn off any room lights you can and try not to allow other light sources to interfere with your measurements.

What you will need:

- A tape measure
- A flashlight
- A Texas Instruments CBL unit with a light sensor probe

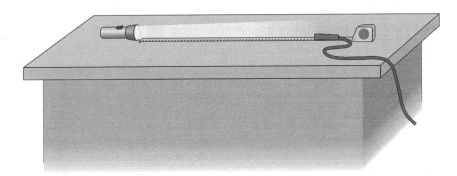

Write up

a. Describe how you collected the data. What was the most important thing to do in order to get good data?

b. Create a scattergram of the data on the calculator or computer and print it out or draw it neatly by hand on graph paper.

c. Use the inverse relationship to find a model to fit the data.

d. What is the variation constant for the light source in your experiment? If someone else in your class also did this experiment, how do the variation constants compare?

e. What is the vertical intercept of your model and what does it represent in this experiment?

f. What is a reasonable domain and range for your model? (Remember to consider any restrictions on the experiment.)

Group Experiment

Three or more people

What you will need:

- A 1-gallon container
- Something to make holes in the container
- Two plugs for the holes
- A stopwatch or timer

DRAIN THAT BUCKET

In this project you will explore the value of things working together. Using a 1-gallon container, you will need to put two different-sized holes in the bottom of the container and plug them closed to start. Fill the container with water and record the time for how long it takes for one of the holes to empty the container. Repeat the experiment with only the second hole open. Finally, repeat the experiment a third time with both holes open at the same time.

Write up

a. Describe the results from each of the three runs of the experiment.

b. Which hole took the least amount of time to drain the container?

c. Did having both holes open make the container drain more quickly? Is this what you expected to happen?

d. Let A be the time it takes hole one to drain the container.

Let B be the time it takes hole two to drain the container.

Using the results from your experiments, fill in the values for the equation

$$\frac{1}{A} + \frac{1}{B} = \frac{1}{t}$$

e. Solve the equation from part d for t. t is the time it should have taken both holes working together to drain the container. How does the value you found for t compare to what you found when you ran the experiment the third time?

EXPLORE YOUR OWN RATIONAL FUNCTION

In this project you are given the task of finding and exploring a real-world situation that you can model using a rational function. You may use the problems in this chapter to get ideas of things to investigate, but your data should not be discussed in this textbook. Some items that you might wish to investigate would be formulas from physics and other sciences. Remember that you can look for inversely proportional relationships as one example.

Write up

Research Project

One or more people

What you will need:

- Find a real-world situation that can be modeled with a rational function.

- You might want to use the Internet or library. Statistical abstracts and some journals and scientific articles are good resources for real-world data.

a. Describe the real-world situation you found and where you found it. Cite any sources you used.

b. Either run your own experiment or find data that can be used to verify the relationship you are investigating.

c. Create a scattergram of the data on the calculator or computer and print it out or draw it neatly by hand on graph paper.

d. Find a model to fit the data.

e. Where are the vertical asymptotes of your model?

f. What is a reasonable domain for your model?

g. Use your model to estimate an output value of your model for an input value that you did not collect in your original data.

h. Use your model to estimate for what input value your model will give you a specific output value that you did not collect in your original data.

Radical Functions

- Identify a radical function from its graph and equation.

- Identitfy square roots and higher roots.

- Use given models to investigate contexts with radical functions.

- Determine the domain and range of radical functions.

- Simplify radical expressions.

- Add and subtract radical expressions.

- Multiply and divide radical expressions.

- Rationalize the denominator of radical expressions.

- Solve radical equations.

- Define imaginary and complex numbers.

- Perform arithmetic operations with complex numbers.

- Find complex solutions to equations.

© Peter Mundy/Alamy

hen riding a bike or driving a car, how fast you are going will determine how sharp of a turn you can make. Cyclists in the Tour de France and other top races can reach downhill speeds in excess of 50 miles per hour. When navigating their way down the back side of a climb, cyclists must be aware of the speed at which they can enter a corner and still maintain control of the bike. In this chapter we will discuss how radical functions relate to a variety of life applications such as skid marks made by a car. Also in one of the chapter projects you will investigate how radical functions relate to the turning radius at different speeds.

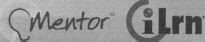

Introduction to Radical Functions

In this chapter we will investigate and work with a new type of function that involves using square roots and other higher roots. We will start by looking at some situations and data that are best modeled by using square roots or other powers that result in radical expressions. The name **radical** is given to the symbol $\sqrt{}$, and the expression inside of the radical is called the **radicand.** The word *radical* is also used to describe any function that uses a radical with variables in the radicand. When we want to represent a root other than a square root we indicate that by using an **index** in the nook of the radical symbol. Whenever the index is higher than 2, the radical is considered a higher root.

DEFINITION
Radical Expression:

$$\sqrt{x} \qquad \sqrt[n]{x}$$

A square root or *n*th root is called a radical expression. In these examples *x* is called the radicand, and *n* is the index. Square roots have an index of 2 but are not written in the nook of the radical.

$$\text{index} \diagdown \sqrt[5]{2x}$$
$$\diagdown \text{radicand}$$

We will give the formal definitions of square roots and higher roots in Section 8.2. To start our investigation of radical functions, let's consider the following data.

MODELING DATA WITH RADICAL FUNCTIONS

 CONCEPT INVESTIGATION I **WHAT SHAPE IS THAT?**

a. Fill in the missing parts to this table of data.

x	0	1	4	9	16	25	36	49	100	144	196
f(x)	0	1	2	3	4	5					

b. Create a scatterplot for this data and describe its shape compared to the shapes of other model types you have learned.

c. Try the following models and describe their fit to these data.

$$y = 0.0666x + 2.305$$
$$y = -0.0003139x^2 + 0.1233x + 1.4236$$

d. Graph $f(x) = \sqrt{x}$ with the data. How well does this fit the data?

e. Change the window to Xmin = −50, Xmax = 250, Ymin = −5, and Ymax = 20. Does this model exist for negative values of *x*? Why or why not?

f. Does this model have negative output values? Why or why not?

Now that we have explored a radical model, let's consider the following applications of some different types of radicals.

EXAMPLE 1 A RADICAL MODEL FROM ZOOLOGY

Allometry is the study of how some aspect of the physiology of a certain species of animal changes in proportion with a change in body size. For example, a study of neotropical butterflies at the Department of Zoology, University of Texas–Austin, studied the relationship of the airspeeds during natural free flight and several characteristics of the butterflies. One relationship studied was between the body mass (in grams) of the butterfly and its mean forward airspeed (in meters per second). A sample of the data collected is given in the table.

Body Mass (in g)	Velocity (in m/s)
0.1	2.75
0.13	3.14
0.16	3.46
0.21	3.99
0.25	4.36
0.3	4.77
0.4	5.49
0.63	6.91

© Kim Garwood/www.neotropicalbutterflies.com

Source: The Journal of Experimental Biology, 191, *125–139 (1994) as found at jeb.biologist.org.*

a. Create a scatterplot of the data in the table.

b. An estimated model for these data is given by the following. Let $V(m)$ be the mean forward airspeed in meters per second for neotropical butterflies and m be the body mass of the butterfly in grams.

$$V(m) = 8.7\sqrt{m}$$

   i. Graph this model with the given data.

   ii. How well does it fit the data?

   iii. Describe the general shape of the graph.

c. Estimate the airspeed of a butterfly with a body mass of 0.5 gram.

d. Give a reasonable domain and range for this model.

e. Use the graph to estimate the body mass of a butterfly that has an airspeed of 5 meters per second.

Solution

a.

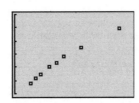

b. i.

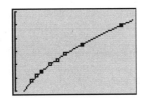

ii. The graph fits the data well and seems to follow the pattern the data are distributed in.

iii. This graph is increasing and is slightly curved and thus is not linear.

c. We are given the body mass of the butterfly, so we can let $m = 0.5$, and we get

$$V(0.5) = 8.7\sqrt{0.5}$$
$$V(0.5) = 6.15$$

Therefore a butterfly with a body mass of 0.5 gram would have a mean forward airspeed of 6.15 meters per second.

d. Because the input variable represents the body mass of a butterfly, we should choose only small positive numbers. Therefore one possible domain could be [0.08, 0.75]. Because the range will be the lowest point to highest point within the domain, we would get a range of [2.46, 7.53].

e. Using the graph and trace, we get the following.

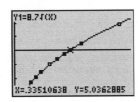

Therefore a butterfly with a body mass of about 0.34 gram will have an airspeed of about 5 meters per second.

The model in Example 1 is a radical function that contains a square root and can be easily evaluated in your calculator. The basic shape of this type of data is an increasing function that is slightly curved. You might note that it is shaped very similarly to a logarithm's graph, the main difference being that it grows more quickly than a log would and does not go into negative output values. In the situation of the butterfly's velocity (airspeed) the outputs would be only positive.

The domain of a radical function that is used in a context can be found the same way that we have done throughout the text. Start by trying to expand the domain beyond the given data avoiding model breakdown and any restrictions that the problem states. The range will then be the lowest to highest output values within that domain. The range of a radical function will again have its output values come from the endpoints of the domain. Let's look at another example that uses a higher root as part of the radical function.

EXAMPLE **2** GROWTH RATE APPLICATION

Pediatricians often use growth charts to compare a child's height and weight to those of the average child of the same age. This helps doctors gauge the growth and development of their patients compared to what is expected in general. The Internet provides parents with several samples of growth charts with heights and weights for girls and boys. Data from one of these charts for girls are given in the table.

Age (in months)	Height (in ft)
4.5	2.0
10.5	2.1667
12	2.4167
24	2.6667
36	2.9167
48	3.1667
60	3.3333

Source: "The Wellness Site," aarogya.com.

a. Create a scatterplot for these data.

b. Let $H(a)$ be the average height of a girl in feet at age a months old.

   i. Graph the function $H(a) = 1.44 \sqrt[5]{a}$ with your data.

   ii. How well does this function fit these data?

c. Estimate the average height of girls who are 18 months old.

d. Estimate the average height of girls who are $2\frac{1}{2}$ years old.

e. What would be a reasonable starting value for the domain of this function?

f. Estimate numerically at what age the average height of a girl would be 2.5 feet.

Solution

a.

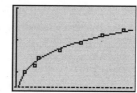

b. i. Note the radical was put into the equation editor using fraction exponents.

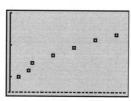 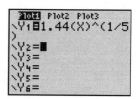

   ii. This function fits the data well and follows the pattern in the data.

c. We are given the age of the girls, so we can substitute $a = 18$ and get

$$H(18) = 1.44\sqrt[5]{18}$$
$$H(18) = 2.5669$$

If we convert the 0.5669 to inches by multiplying by 12, we get $0.5669 \cdot 12 = 6.8$ inches. Therefore the average height of an 18-month-old girls is about 2 feet 7 inches.

d. We are told that the girls are $2\frac{1}{2}$ years old, so we can substitute $a = 30$ months and get

$$H(30) = 1.44\sqrt[5]{30}$$
$$H(30) = 2.8431$$

If we convert the 0.8431 into inches, we get $0.8431 \cdot 12 = 10.1$ inches. Therefore the average height of $2\frac{1}{2}$ year old girls is about 2 feet 10 inches.

e. Because we are talking about the average height of girls at age a, the domain will be the ages of girls in months. Therefore negative values would not make sense, and $a = 0$ would result in an average height of zero feet, so that also is model breakdown. We should probably start this domain at about 3 months of age. This gives us a reasonable height of about 21.5 inches.

f. Using the table and starting at 12 months, we get the following.

X	Y1
12	2.367
13	2.4052
14	2.4411
15	2.475
16	2.5072
17	2.5378

X=

Therefore the average height of 16 month old girls is 2.5 feet.

Both of these examples require you to evaluate the function for a given value of the input variable, which is easily done on the calculator. We have also solved these functions using a graph or table. To solve these types of models algebraically for a missing value of the input variable, we will need several other skills that we will be learning in the next few sections. In a later section of this chapter we will come back to some of these examples and solve them algebraically for the input variables.

EXAMPLE ② PRACTICE PROBLEM

a. Given the function $f(x) = 7.2\sqrt[3]{x}$ find the following.

    i. $f(20)$ **ii.** Estimate numerically when $f(x) = 34$.

b. Given the graph of the function $h(x)$ find the following.

i.   Estimate $h(2)$.   **ii.**   Estimate $h(x) = -8$.

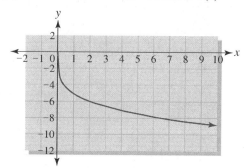

DOMAIN AND RANGE OF RADICAL FUNCTIONS

Let's look at the graphs of several radical functions and consider the domain and range of those functions in problems without a context.

 CONCEPT INVESTIGATION 2 ODD OR EVEN--DOES IT MATTER?

Set your graphing calculator's window to Xmin $= -10$, Xmax $= 10$, Ymin $= -3.5$, and Ymax $= 3.5$.

a.   Graph the following functions on your graphing calculator.

  i. $f(x) = \sqrt{x}$.

  ii. $g(x) = \sqrt[4]{x}$. Remember to enter the index with fraction exponents.

  iii. $h(x) = \sqrt[6]{x}$.

  Describe the shape of these graphs.
    How does the graph change as you take higher roots?
    What appear to be the domain and range for these functions?

b.   Graph the following functions on your graphing calculator.

  i. $f(x) = \sqrt[3]{x}$.   **ii.** $g(x) = \sqrt[5]{x}$.   **iii.** $h(x) = \sqrt[7]{x}$.

  Describe the shape of these graphs.
  How does the graph change as you get higher roots?
  What appear to be the domain and range for these functions?
  What is the difference between the radical functions in part a and the radical functions in part b?

As you can see from this concept investigation, whether the root is odd or even makes a difference in the shape of the graph as well as what its domain and range are. You should note that, in the real number system, even roots cannot have negative numbers under the radical and thus have only values that make the radicand nonnegative as part of their domain. Odd roots do not have this restriction because negatives are possible under an odd root and thus typically have all real numbers as a domain. We will study what happens when negatives are under a square root in the last section of this chapter.

GRAPHS OF RADICAL FUNCTIONS

Even roots.

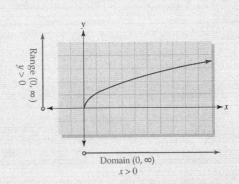

$$f(x) = \sqrt[n]{x}$$

n even

Even roots must have inputs that keep the radicand positive. This will restrict the domain and range.

Odd roots.

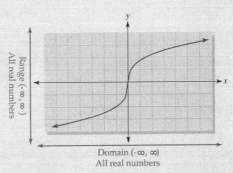

$$f(x) = \sqrt[n]{x}$$

n odd

Odd roots can have any real number radicand, so the domain and range will be all real numbers.

EXAMPLE   3   DOMAIN AND RANGE OF RADICAL FUNCTIONS

Give the domain and range of the following radical functions.

a. $f(x) = \sqrt{x + 3}$.

b. $g(x) = \sqrt{x-9}$.

c.

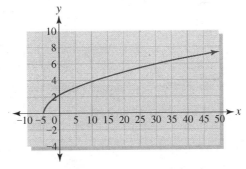

d. $f(x) = \sqrt[3]{x-7}$.

e.

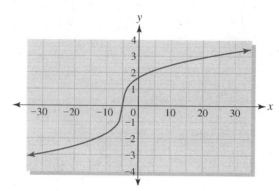

f. $h(x) = \sqrt{-x}$.

g.

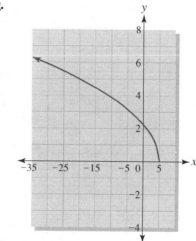

Solution

a. The domain can include only values that make the radicand nonnegative, so we solve the following inequality

$$x + 3 \geq 0$$
$$x \geq -3$$

Therefore the domain of this function is $x \geq -3$. The lowest point on the graph is zero. The graph continues upward to infinity, so we get a range of $[0, \infty)$. We can confirm this domain and range by looking at its graph on the calculator.

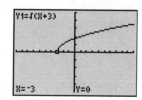

b. The domain can include only values that make the radicand nonnegative, so we need to solve the following inequality

$$x - 9 \geq 0$$
$$x \geq 9$$

Therefore the domain of this function is $x \geq 9$. The lowest point on the graph of this function is $(9, 0)$, so the range will start at zero. The function continues up to infinity, so we get a range of $[0, \infty)$. We can confirm this domain and range by looking at its graph on the calculator.

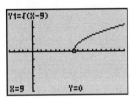

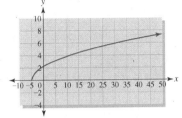

c. By looking at this graph, we can see that the input values that are being used by this function start at $x = -5$. Therefore the domain of this function should be $x \geq -5$. The range will start at zero and go up to infinity, giving us a range of $[0, \infty)$.

d. This function is cubic, which means that is has an odd root. The radicand may be either positive or negative, so the domain would include all real numbers. The range will also be all real numbers because the graph goes down to negative infinity and up to positive infinity. Looking at the graph of this function confirms this.

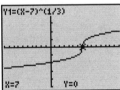

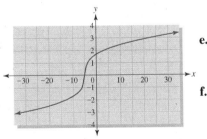

e. This graph uses all the input values and appears to be an odd root, so the domain of this function will be all real numbers. The range is also all real numbers because the graph goes down to negative infinity and up to positive infinity.

f. Because this is an even root, the radicand must be nonnegative. Although the radicand looks negative, if x is a negative number, then $-x$ will be a positive number. We solve the following inequality.

$$-x \geq 0$$
$$\frac{-x}{-1} \leq \frac{0}{-1}$$
$$x \leq 0$$

To isolate x, we divide both sides by -1. Remember that we need to flip the inequality symbol whenever we multiply or divide both sides by a negative number.

Therefore the domain of this function is $x \leq 0$. The lowest point on the graph of this function is $(0, 0)$, so the range will start at zero. The function continues up to infinity, so we get a range of $[0, \infty)$. We can confirm this domain and range by looking at its graph on the calculator.

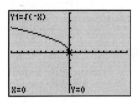

g. From this graph we can see that only input values less than or equal to 5 are being used. The domain will be $x \leq 5$. The range will be $[0, \infty)$.

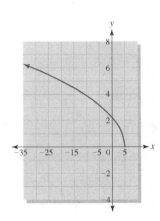

EXAMPLE ③ PRACTICE PROBLEM

Give the domain and range of the following radical functions.

a. $f(x) = \sqrt{x + 8}$

b. $g(x) = \sqrt{-x + 3}$

c. $h(x) = \sqrt[5]{x - 9}$

d.

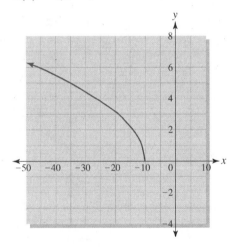

8.1 Exercises

1. Jim Bob's Heat Source manufactures h hundred thousand space heaters per month. Jim Bob has $3.2 million in monthly fixed costs to run the plant. The following data have been collected on the cost of various numbers of heaters produced per month.

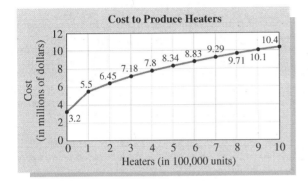

Cost to Produce Heaters

Height (in ft)	Drop Time (in sec)
4	0.49
6	0.6
11.5	0.82
18.5	1.05
23	1.17
37	1.48
54.5	1.8

a. Create a scatterplot for these data. Does a radical function seem appropriate for this model?

b. Let $C(h)$ be the cost in millions of dollars to produce h hundred thousand heaters per month. Graph the following function with the data.

$$C(h) = 2.29\sqrt{h} + 3.2$$

How well does this model fit the data?

c. Use the model to estimate the cost to produce 550,000 heaters per month.

d. Use the model to estimate the cost to produce 1,500,000 heaters per month.

e. Give a reasonable domain and range for this model.

2. The physics department is doing an experiment testing the time it takes for an object to fall from different heights. One group collected the following data.

a. Let $T(h)$ be the drop time in seconds for an object that is dropped from h feet. Plot the data and the model

$$T(h) = 0.243\sqrt{h}$$

b. Use the model to determine the drop time for an object dropped from 15 feet.

c. Use the model to determine the drop time for an object that is dropped from 100 feet.

d. Determine a reasonable domain and range for this model.

e. Estimate numerically what height an object must be dropped from to have a drop time of 2 seconds.

3. In a study of basilisk lizards, a biologist found that there were several allometric relationships between the body mass of a lizard and different characteristics of its body. One such relationship is demonstrated by the following data and model.

Body Mass (in g)	Leg Length (in m)
2.5	0.022
15	0.041
34.9	0.054
41.7	0.057
72.1	0.069
90.3	0.074
140.6	0.086
164.2	0.09

Source: Data and model derived from Size-Dependence of Water Running Ability in Basilisk Lizards *by Blasheen and McMahon. The Journal of Experimental Biology 199, 2611–2618, 1996.*

© Buddy Mays/CORBIS

Let $L(M)$ represent the leg length in meters of a basilisk lizard with a body mass of M grams.

$$L(M) = 0.0165 \sqrt[3]{M}$$

a. How well does this model fit the data given?

b. Estimate the leg length of a basilisk lizard that has a body mass of 50 grams.

c. Estimate the leg length of a basilisk lizard that has a body mass of 100 grams.

d. Estimate graphically the body mass of a basilisk lizard that has a leg length of 0.1 m.

4. Dr. Marina Silva, in the *Journal of Mammalogy* (1998), gives approximately the following formula for the body length of mammals based on the mammal's body mass.

$$L(M) = 0.330 \sqrt[3]{M}$$

where $L(M)$ represents the body length in meters of a mammal with a body mass of M kilograms.

a. Use this model to estimate the body length of a mammal with a body mass of 4.6 kg.

b. Use this model to estimate the body length of a mammal with a body mass of 25 kg.

c. If this model is valid for mammals with body masses between 0.01 and 250 kg, what is this model's range?

d. Estimate numerically the body mass of a mammal with a body length of 1 m.

5. The butterfly study from Example 1 also developed the following formula for the relationship between the thoracic (middle body portion) mass of the butterfly and its mean forward airspeed.

$$V(m) = 17.6 \sqrt[5]{m^3}$$

where $V(m)$ represents the mean forward airspeed in meters per second of a butterfly with a thoracic mass of m grams..

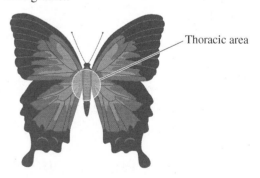

Thoracic area

a. Use the model to estimate the airspeed of a butterfly with a thoracic mass of 0.05 gram.

b. Use the model to estimate the airspeed of a butterfly with a thoracic mass of 0.12 gram.

c. If this model is valid for butterflies with a thoracic mass between 0.001 and 0.6 gram, determine the range of this model

6. The average height of boys for various ages are given in the table.

Age (in months)	Average Height of Boys (in feet)
4.5	2.05
10.5	2.3
12	2.45
24	2.75
36	2.95
48	3.2

Source: "The Wellness Site," aarogya.com.

   a. Let $H(a)$ be the average height in feet of boys at a months old. The height can be modeled by

   $$H(a) = 1.46 \sqrt[5]{a}$$

   Graph the data and the model together. How well does this model fit the data?

   b. Estimate the average height for boys who are 18 months old.

   c. Estimate the average height for boys who are $2\frac{1}{2}$ years old.

   d. How do your answers in parts b and c compare to the average heights of girls found in Example 2?

   e. Give a reasonable domain and range for this model.

7.

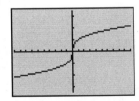

   a. Is this the graph of an odd or even root?
   b. Give the domain of this function.
   c. Give the range of this function.

8.

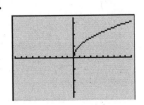

   a. Is this the graph of an odd or even root?
   b. Give the domain of this function.
   c. Give the range of this function.

9.

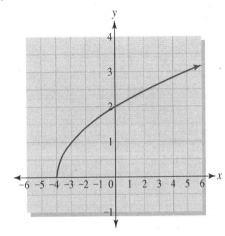

   a. Is this the graph of an odd or even root?
   b. Give the domain of this function.
   c. Give the range of this function.

10.

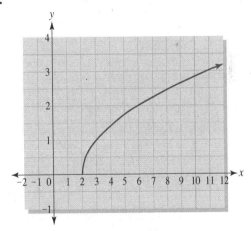

   a. Is this the graph of an odd or even root?
   b. Give the domain of this function.
   c. Give the range of this function.

11.

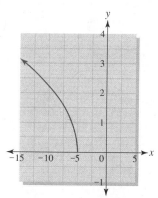

a. Is this the graph of an odd or even root?

b. Give the domain of this function.

c. Give the range of this function.

12.

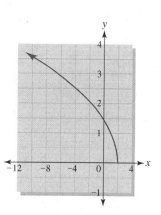

a. Is this the graph of an odd or even root?

b. Give the domain of this function.

c. Give the range of this function.

13.

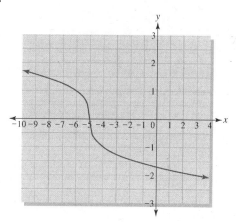

a. Is this the graph of an odd or even root?

b. Give the domain of this function.

c. Give the range of this function.

14.

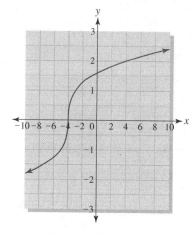

a. Is this the graph of an odd or even root?

b. Give the domain of this function.

c. Give the range of this function.

15.

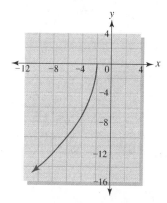

a. Is this the graph of an odd or even root?

b. Give the domain of this function.

c. Give the range of this function.

16.

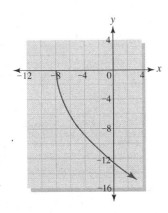

a. Is this the graph of an odd or even root?

b. Give the domain of this function.

c. Give the range of this function.

For Exercises 17 through 32 give the domain and range of the given function.

17. $f(x) = 4\sqrt[3]{x}$

18. $g(m) = -3\sqrt[5]{m}$

19. $f(x) = 5\sqrt{x}$

20. $g(x) = -7\sqrt{x}$

21. $h(x) = 0.7\sqrt[9]{x}$

22. $f(x) = -\dfrac{5}{8}\sqrt[7]{x}$

23. $h(x) = 0.24\sqrt[4]{x}$

24. $f(x) = -\dfrac{2}{3}\sqrt[8]{x}$

25. $f(x) = \sqrt{x + 12}$

26. $g(x) = \sqrt{x - 6}$

27. $f(x) = \sqrt[3]{x + 15}$

28. $g(x) = \sqrt[3]{-x + 12}$

29. $h(x) = \sqrt{-x + 3}$

30. $f(x) = \sqrt{x + 8}$

31. $g(x) = \sqrt{-x + 10}$

32. $h(x) = \sqrt{-x - 11}$

SECTION 8.2

Simplifying, Adding, and Subtracting Radicals

SQUARE ROOTS AND HiGHER ROOTS

Radicals include square roots and other higher roots such as cube roots and fourth roots. To solve problems that involve radicals, you should know some basic rules for doing arithmetic operations. In this section you will learn to simplify radical expressions and perform basic operations with radicals.

Let's start with a basic definition of a square root.

DEFINITION

Square root:

b is the square root of a if $b^2 = a$.

$$\sqrt{a} = b$$

Using rational exponents we get:

$$\sqrt{a} = \sqrt{b^2} = (b^2)^{\frac{1}{2}} = b$$

$$\sqrt{25} = \sqrt{5^2} = (5^2)^{\frac{1}{2}} = 5$$

Although you are probably more familiar with square roots, higher roots actually have a very similar definition.

Recall from Section 3.1 that a rational exponent is the same as a radical.

$$x^{\frac{1}{2}} = \sqrt{x} \qquad x^{\frac{1}{n}} = \sqrt[n]{x}$$

If you think of radicals as rational exponents, you can perform operations on radicals using the rules for exponents. The most important rules for exponents to remember are the following.

Product Rule:

$$x^m x^n = x^{m+n}$$

$$x^{\frac{1}{5}} x^{\frac{2}{3}} = x^{\frac{2}{15}}$$

Quotient Rule:

$$\frac{x^m}{x^n} = x^{m-n}$$

$$\frac{x^4}{x^{3\frac{1}{3}}} = x^{\frac{2}{3}}$$

Power Rule:

$$(x^m)^n = x^{mn}$$

$$(x^4)^7 = x^{28}$$

Distributing Exponents:

$$(xy)^m = x^m y^m \qquad \left(\frac{x}{y}\right)^m = \frac{x^m}{y^m}$$

$$(xy)^5 = x^5 y^5 \qquad \left(\frac{x}{y}\right)^4 = \frac{x^4}{y^4}$$

Using these rules is one option for simplifying and performing most operations on radicals.

DEFINITION

nth root:

b is the nth root of a if $b^n = a$.

$$\sqrt[n]{a} = b$$

n is called the index of the radical, a is called the radicand, and b is called the root.

Using rational exponents, we get

$$\sqrt[n]{a} = \sqrt[n]{b^n} = (b^n)^{\frac{1}{n}} = b$$

$$\sqrt[3]{8} = \sqrt[3]{2^3} = (2^3)^{\frac{1}{3}} = 2$$

$$\sqrt[4]{81} = \sqrt[4]{3^4} = (3^4)^{\frac{1}{4}} = 3$$

$$\sqrt[3]{-125} = \sqrt[3]{(-5)^3} = ((-5)^3)^{\frac{1}{3}} = -5$$

When working with radicals, you might notice that square roots always seem to be positive. This is true only by the choice of mathematicians. In general, the square root would have both a positive and negative answer.

$$\sqrt{25} = \begin{cases} \sqrt{5^2} = 5 \\ \sqrt{(-5)^2} = -5 \end{cases}$$

Because having two possible answers will cause many problems with consistency, mathematicians agree to use the **principal root,** which is the positive root only. If a negative root is needed than we use a negative sign outside of the square root to designate the negative root

$$-\sqrt{100} = -10 \qquad \text{but} \qquad \sqrt{100} = 10$$

You should note that a root with an odd index can have a negative radicand and thus will have a negative root. If a root with an even index has a negative radicand, the root will not be a real number. We will study square roots with negative radicands in Section 6.5.

EXAMPLE   **1**   EVALUATING RADICALS

Evaluate the following radicals.

a. $\sqrt{36}$ **b.** $\sqrt[3]{8}$

c. $\sqrt[3]{-27}$ **d.** $\sqrt[4]{625}$

Solution

a. $\sqrt{36} = \sqrt{6^2} = 6$ **b.** $\sqrt[3]{8} = \sqrt[3]{2^3} = 2$

c. $\sqrt[3]{-27} = \sqrt[3]{(-3)^3} = -3$ **d.** $\sqrt[4]{625} = \sqrt[4]{5^4} = 5$

EXAMPLE ① PRACTICE PROBLEM

Evaluate the following radicals.

a. $\sqrt{49}$ **b.** $\sqrt{144}$

c. $\sqrt[3]{-512}$ **d.** $\sqrt[4]{10000}$

SIMPLIFYING RADICALS

When simplifying radical expressions, you are looking for any factors of the radicand that can be pulled out of the radical. Anything that cannot come out of the radical can be left inside as a new reduced radicand. Some radicals will contain variables and thus will only be able to be simplified when the value of those variables is known. Whenever we are given a radical with a variable in the radicand, we will assume that the variable represents only nonnegative values. Then we will not have to worry about inputs that are not in the domain. In working with variables in a radicand, using the rules for exponents will simplify the process.

STEPS TO SIMPLIFY A RADICAL EXPRESSION

1. Break the radicand into factors that are raised to the same power as the index of the radical.

2. Reduce the radicals.

3. Simplify by multiplying any remaining radicands together and multiplying anything that has been taken out of the radicals.

STEPS TO SIMPLIFY A RADICAL EXPRESSION USING THE RULES FOR EXPONENTS

1. Rewrite the radical in exponential form.

2. Factor the constant.

3. Distribute the exponent.

4. Write exponents as mixed numbers. (if needed)

5. Separate the whole and fraction exponent parts and rewrite in radical form.

6. Multiply any remaining constant parts together.

EXAMPLE ② SIMPLIFYING RADICALS

Simplify the following radicals.

a. $\sqrt{25x^2}$ **b.** $\sqrt[3]{8m^3n^6}$

c. $\sqrt{15x^3y^6}$ **d.** $\sqrt[5]{96a^2b^8c^{10}}$

Solution

a. $\sqrt{25x^2}$

$\sqrt{5^2}\sqrt{x^2}$ Break the radicand into factors that are squared.

$5x$ Reduce.

b. $\sqrt[3]{8m^3n^6}$ Because this is a cube root, we will look for factors that are cubed.

$\sqrt[3]{2^3}\sqrt[3]{m^3}\sqrt[3]{n^3}\sqrt[3]{n^3}$ Break the radicand into factors that are cubed.

$2mnn$ Reduce.

$2mn^2$ Simplify.

c. We will do this problem two ways. First in a similar way to parts a and b.

$$\sqrt{15x^3y^6}$$

$\sqrt{x^2}\sqrt{y^2}\sqrt{y^2}\sqrt{y^2}\sqrt{3}\sqrt{5}\sqrt{x}$ Break the radicand into factors that are squared.

$xyyy\sqrt{3}\sqrt{5}\sqrt{x}$ Reduce and group together remaining radicals.

$xy^3\sqrt{15x}$ Simplify.

Now we will do it again, using the rules for exponents.

$\sqrt{15x^3y^6} = (15x^3y^6)^{\frac{1}{2}}$ Rewrite in exponential form.

$= (3 \cdot 5x^3y^6)^{\frac{1}{2}}$ Factor the constant.

$= 3^{\frac{1}{2}} \cdot 5^{\frac{1}{2}}x^{\frac{3}{2}}y^3$ Distribute the exponents.

$= 3^{\frac{1}{2}} \cdot 5^{\frac{1}{2}}x^{1\frac{1}{2}}y^3$ Write exponents as mixed numbers if needed.

$= xy^3\sqrt{15x}$ Separate the whole and fraction exponent parts and rewrite in radical form. Multiply the constants together.

d. We will do this problem using the rules for exponents.

$\sqrt[5]{96a^2b^8c^{10}} = (96a^2b^8c^{10})^{\frac{1}{5}}$ Rewrite in exponential form.

$= (32 \cdot 3a^2b^8c^{10})^{\frac{1}{5}}$ Factor the constant.

$= (2^5 \cdot 3a^2b^8c^{10})^{\frac{1}{5}}$

$= 2 \cdot 3^{\frac{1}{5}}a^{\frac{2}{5}}b^{\frac{8}{5}}c^2$ Distribute the exponents.

$= 2 \cdot 3^{\frac{1}{5}}a^{\frac{2}{5}}b^{1\frac{3}{5}}c^2$ Write exponents as mixed numbers.

$= 2bc^2\sqrt[5]{3a^2b^3}$ Separate the whole and fraction exponent parts and rewrite in radical form.

You should note that all of these problems could be reduced by using either method. It is often best if you find which of these methods you like and use that method consistently. As you simplify more radicals, you might find that you do not need to write out all these steps. You should use caution when working with fraction exponents so that you do not make mistakes.

Skill Connection

Some students find it easiest to change radicals to rational exponents and then use these rules to simplify the problem. When a problem is completed, you should change the answers back into radical form.

EXAMPLE ② PRACTICE PROBLEM

Simplify the following radicals.

a. $\sqrt{16a^2b^4}$

b. $\sqrt{180x^2y^3}$

c. $\sqrt[3]{108m^5n^6}$

d. $\sqrt[9]{x^{36}y^{18}z^{11}}$

ADDING AND SUBTRACTING RADICALS

Adding and subtracting radical expressions is basically the same as adding and subtracting like terms in a polynomial. You can add or subtract only two radical expressions that have the same index, the same variables and exponents outside the radical and the same radicand inside. This can also be looked at as a form of factoring.

$$5\sqrt{3x} + 7\sqrt{3x}$$ Factor out the $\sqrt{3x}$ from both terms.

$$\sqrt{3x}(5 + 7)$$ Combine the constants, and you have combined

$$12\sqrt{3x}$$ like terms.

DEFINITION

Like Radical Expression:

Radical expressions are like if they have the same index, the same radicand and the same variables and exponents outside the radical.

STEPS TO ADDING OR SUBTRACTING RADICAL EXPRESSIONS

1. Simplify each radical.

2. Add or subtract any like radical expressions by adding or subtracting the coefficients in front of the radical and keeping the radicand the same.

EXAMPLE ③ ADDING AND SUBTRACTING RADICALS

Add or subtract the following expressions.

a. $3\sqrt{5} + 7\sqrt{5}$

b. $2x\sqrt{10} - 5x\sqrt{10}$

c. $\sqrt{25x^3yz^5} + 2x\sqrt{xyz^5}$

d. $5\sqrt{x} + 2\sqrt{3x} + 4\sqrt{x} - 7\sqrt{3x}$

Solution

a. $3\sqrt{5} + 7\sqrt{5} = 10\sqrt{5}$ Combine like terms.

b. $2x\sqrt{10} - 5x\sqrt{10} = -3x\sqrt{10}$ Combine like terms.

c. $\sqrt{25x^3yz^5} + 2x\sqrt{xyz^5} = 5xz^2\sqrt{xyz} + 2xz^2\sqrt{xyz}$ Simplify each radical.

$$= 7xz^2\sqrt{xyz}$$ Combine like terms.

d. $5\sqrt{x} + 2\sqrt{3x} + 4\sqrt{x} - 7\sqrt{3x} =$

Combine like terms.

$$= (5\sqrt{x} + 4\sqrt{x}) + (2\sqrt{3x} - 7\sqrt{3x})$$
$$= 9\sqrt{x} - 5\sqrt{3x}$$

EXAMPLE ③ PRACTICE PROBLEM

Simplify the following by adding or subtracting the like expressions.

a. $2\sqrt{3} - 7\sqrt{3}$

b. $4\sqrt{50} + 7\sqrt{2}$

c. $3x\sqrt{5y^3} + 2xy\sqrt{5y}$

d. $\sqrt{5x} + 2\sqrt{2x} - 6\sqrt{2x} + 4\sqrt{5x}$

When combining radical expressions, you only combine like expressions. This will include only combining similar radicals. For instance, you should not add or subtract a square root with a cube root or any other higher root.

EXAMPLE ④ ADDING AND SUBTRACTING RADICALS

Simplify the following by adding or subtracting the like expressions.

a. $5\sqrt{x} + 2\sqrt[3]{x} - 2\sqrt{x} + 3\sqrt[3]{x}$

b. $\sqrt[3]{8x^5y^2} + 5x\sqrt[4]{x^2y^2} + 7x\sqrt[3]{x^2y^2}$

Solution

a. $5\sqrt{x} + 2\sqrt[3]{x} - 2\sqrt{x} + 3\sqrt[3]{x} = 3\sqrt{x} + 5\sqrt[3]{x}$

Combine like expressions

b. $\sqrt[3]{8x^5y^2} + 5x\sqrt[4]{x^2y^2} + 7x\sqrt[3]{x^2y^2} = 2x\sqrt[3]{x^2y^2} + 5x\sqrt[4]{x^2y^2} + 7x\sqrt[3]{x^2y^2}$

$$= 9x\sqrt[3]{x^2y^2} + 5x\sqrt[4]{x^2y^2}$$

EXAMPLE ④ PRACTICE PROBLEM

Simplify the following by adding or subtracting the like expressions.

a. $12\sqrt{2m} + 5\sqrt[3]{7m} - 4\sqrt{2m} + 3\sqrt[3]{7m}$

b. $\sqrt[5]{32x^3y} + 5\sqrt[4]{xy^2} + 6\sqrt[5]{x^3y}$

8.2 Exercises

For Exercises 1 through 20, simplify the given radical expression.

1. $\sqrt{100}$

2. $\sqrt{49x^2}$

3. $\sqrt[3]{125y^6}$

4. $\sqrt[4]{16a^4b^{12}}$

5. $\sqrt[3]{-8}$

6. $\sqrt[5]{-32c^5d^{10}}$

7. $\sqrt{196m^4n^2}$

8. $\sqrt{50x^2y}$

9. $\sqrt{36a^3}$

10. $\sqrt{24w^4z^5}$

11. $\sqrt{1296a^5b^8c^{15}}$

12. $\sqrt{6480x^3y^5z^8}$

13. $\sqrt[3]{8x^3y^6}$

14. $\sqrt[3]{40a^3b^5}$

15. $\sqrt[4]{48a^3b^8c^{21}}$

16. $\sqrt[5]{32m^5n^{10}}$

17. $\sqrt[5]{243s^3t^5u^{10}}$

18. $\sqrt[3]{3888x^3y^5z^8}$

19. $\sqrt[3]{-27a^3b^6}$

20. $\sqrt[3]{-960m^3n^4p^5}$

For Exercises 21 through 38, perform the indicated operation and simplify.

21. $\sqrt{5x} + 3\sqrt{5x}$

22. $2x\sqrt{3} + 7x\sqrt{3} + 4x\sqrt{3}$

23. $2\sqrt{4x^2y} - 3x\sqrt{y}$

24. $5\sqrt{49m^2n^5} - 3m\sqrt{16n^5}$

25. $\sqrt[3]{8a} + 4\sqrt[3]{a}$

26. $3\sqrt[5]{64a^5b^{10}c^3} - 10ab\sqrt[5]{2b^5c^3}$

27. $\sqrt[5]{10m^3n^5} + 4n\sqrt[5]{10m^3}$

28. $\sqrt[4]{16x^5y^8} + 7xy^2\sqrt[4]{81x}$

29. $7\sqrt[3]{125x^6y^9z^2} - 9x^2y\sqrt[3]{216y^3z^2}$

30. $2\sqrt[3]{1080a^3b^6c^9} - 8abc\sqrt[3]{625b^3c^6}$

31. $5\sqrt{2x} + 7x\sqrt{2} - 12\sqrt{2x} + 3x\sqrt{2}$

32. $4\sqrt{18} + 5\sqrt{2} - 8\sqrt{75} + 3\sqrt{48}$

33. $3xy\sqrt{9z^5} + 7xz^2\sqrt{yz} - 2xyz^2\sqrt{z}$

34. $5m^2n\sqrt{12mn^2} + 7mn^2\sqrt{3mn} - 10\sqrt{243m^5n^4}$

35. $5\sqrt{2x} + 4\sqrt[3]{2x} + 7\sqrt{2x}$

36. $5x\sqrt{14mn} + 2x\sqrt[4]{14mn} - 7x\sqrt[4]{14mn}$

37. $\sqrt[3]{7xy} + \sqrt[5]{7xy} - \sqrt[3]{448xy} + \sqrt[5]{224xy}$

38. $\sqrt{27x^5y^3} + \sqrt[3]{27x^5y^3} - 3xy\sqrt[3]{x^2} + xy\sqrt{3x^3y}$

Multiplying and Dividing Radicals

MULTIPLYING RADICALS

Multiplying radical expressions is not a difficult process. Multiplying radicals is done using the exponent rules. When you multiply two radical expressions with the same index, you multiply the radicands (insides) together. After you multiply the radicands, you should simplify the result if possible. Remember that you do not need like expressions to multiply, which is not teh same as teh opperations of addition and subtraction.

MULTIPLYING RADICALS

$$\sqrt[n]{a} \cdot \sqrt[n]{b} = \sqrt[n]{ab}$$

Using exponent rules:

$$\sqrt[n]{a} \cdot \sqrt[n]{b} = a^{\frac{1}{n}} \cdot b^{\frac{1}{n}} = (ab)^{\frac{1}{n}} = \sqrt[n]{ab}$$

When multiplying radicals, multiply the radicands together. The indices of each radical must be the same.

$$\sqrt{3} \cdot \sqrt{7} = \sqrt{21}$$

$$\sqrt[4]{5x} \cdot \sqrt[4]{6xy} = \sqrt[4]{30x^2y}$$

EXAMPLE 1 MULTIPLYING RADICALS

Multiply the following and simplify the result.

a. $\sqrt{5} \cdot \sqrt{2}$

b. $\sqrt{2x} \cdot \sqrt{10y}$

c. $\sqrt{2a} \cdot \sqrt{18a}$

d. $\sqrt[3]{4m^2} \cdot \sqrt[3]{5m}$

Solution

a. $\sqrt{5} \cdot \sqrt{2} = \sqrt{5 \cdot 2}$
$$= \sqrt{10}$$

b. $\sqrt{2x} \cdot \sqrt{10y} = \sqrt{20xy}$
$$= \sqrt{4}\sqrt{5xy} \quad \text{Simplify.}$$
$$= 2\sqrt{5xy}$$

c. $\sqrt{2a} \cdot \sqrt{18a} = \sqrt{36a^2}$
$$= 6a \quad \text{Simplify.}$$

d. $\sqrt[3]{4m^2} \cdot \sqrt[3]{5m} = \sqrt[3]{20m^3} = m\sqrt[3]{20}$

EXAMPLE 1 PRACTICE PROBLEM

Multiply the following and simplify the result.

a. $\sqrt[3]{3a} \cdot \sqrt[3]{5bc}$

b. $\sqrt{5m} \cdot \sqrt{20m^3}$

c. $\sqrt{3y} \cdot \sqrt{12y}$

If the expressions are more complicated, you will multiply the constants and variables outside of the radicals together and multiply the constants and variables inside the radicals together. Once you have multiplied each of these together, you should simplify the result. Some students find it easiest to simply each radical first, then multiply them together, and simplify again if necessary.

EXAMPLE 2 MULTIPLYING RADICALS

Multiply the following and simplify the result.

a. $5\sqrt{2x} \cdot 3\sqrt{7}$

b. $2x\sqrt{3y} \cdot 5\sqrt{12y}$

c. $5a^2b\sqrt{3ab^3} \cdot 2ab^2\sqrt{21a^2b}$

d. $2mn^2\sqrt[4]{m^2n} \cdot 5mn^3\sqrt[4]{mn^9}$

Solution

a. $5\sqrt{2x} \cdot 3\sqrt{7} = 15\sqrt{14x}$

b. $2x\sqrt{3y} \cdot 5\sqrt{12y} = 10x\sqrt{36y^2}$ **Multiply the insides and outsides.**
$$= 10x \cdot 6y \quad \text{Simplify the radical.}$$
$$= 60xy$$

c. $5a^2b\sqrt{3ab^3} \cdot 2ab^2\sqrt{21a^2b} = 10a^3b^3\sqrt{63a^3b^4}$

$$= 10a^3b^3\sqrt{3^2 \cdot 7a^3b^4}$$

$$= 10a^3b^3\sqrt{3^2a^2b^4}\sqrt{7a}$$

$$= 10a^3b^3(3ab^2)\sqrt{7a}$$

$$= 30a^4b^5\sqrt{7a}$$

Break the radicand into factors that are multiples of 2.
Reduce.
Simplify.

d. $2mn^2\sqrt[4]{m^2n} \cdot 5mn^3\sqrt[4]{mn^9} = 10m^2n^5\sqrt[4]{m^3n^{10}}$

$$= 10m^2n^5\sqrt[4]{m^3n^{10}}$$

$$= 10m^2n^5\sqrt[4]{n^8}\sqrt[4]{m^3n^2}$$

$$= 10m^2n^5(n^2)\sqrt[4]{m^3n^2}$$

$$= 10m^2n^7\sqrt[4]{m^3n^2}$$

Break the radicand into factors that are multiples of 4.
Reduce.
Simplify.

EXAMPLE ② PRACTICE PROBLEM

Multiply the following and simplify the result.

a. $4\sqrt{3a} \cdot 6\sqrt{5b}$

b. $3m\sqrt{14n} \cdot 5\sqrt{21mn}$

c. $7x^2y^3\sqrt[3]{x^2y^5} \cdot 4xy\sqrt[3]{xy^2}$

DIVIDING RADICALS AND RATIONALIZING THE DENOMINATOR

Division inside a radical can be simplified in the same way that a fraction would be reduced if it were by itself. This follows from the rules for distributing exponents.

$$\sqrt{\frac{50}{2}} = \sqrt{25} = 5$$

$$\sqrt{\frac{50}{2}} = \left(\frac{50}{2}\right)^{\frac{1}{2}} = 25^{\frac{1}{2}} = 5$$

We can use these rules to simplify some radical expressions that have fractions in them. Please note that you can simplify only fractions that are both inside a radical or both outside the radical. You cannot cancel something that is inside the radical with something that is outside of a radical.

RADICALS WITH FRACTIONS

$$\sqrt[n]{\frac{a}{b}} = \frac{\sqrt[n]{a}}{\sqrt[n]{b}}$$

Using exponent rules:

$$\sqrt[n]{\frac{a}{b}} = \left(\frac{a}{b}\right)^{\frac{1}{n}} = \frac{a^{\frac{1}{n}}}{b^{\frac{1}{n}}} = \frac{\sqrt[n]{a}}{\sqrt[n]{b}}$$

A fraction inside a radical can be made into a fraction with separate radicals in the numerator and denominator of the fraction. This rule is often used in both directions.

EXAMPLE 3 SIMPLIFYING RADICALS WITH DIVISION

Simplify the following radicals.

a. $\sqrt{\dfrac{100}{25}}$ **b.** $\sqrt{\dfrac{49}{36}}$

c. $\sqrt{\dfrac{400x^3y}{4xy^5}}$ **d.** $\sqrt[3]{\dfrac{a^5b^2}{a^2b}}$

Solution

a. $\sqrt{\dfrac{100}{25}} = \sqrt{4}$

$= 2$

b. $\sqrt{\dfrac{49}{36}} = \dfrac{\sqrt{49}}{\sqrt{36}}$

$= \dfrac{7}{6}$

c. $\sqrt{\dfrac{400x^3y}{4xy^5}} = \sqrt{\dfrac{100x^2}{y^4}}$ **Reduce the fraction.**

$= \dfrac{\sqrt{100x^2}}{\sqrt{y^4}}$ **Separate the radical and reduce.**

$= \dfrac{10x}{y^2}$

d. $\sqrt[3]{\dfrac{a^5b^2}{a^2b}} = \sqrt[3]{a^3b}$ **Reduce the fraction.**

$= a\sqrt[3]{b}$ **Reduce the radical.**

EXAMPLE ③ PRACTICE PROBLEM

Simplify the following radicals.

a. $\sqrt{\dfrac{42}{7}}$ **b.** $\sqrt{\dfrac{50}{18}}$ **c.** $\sqrt[4]{\dfrac{144m^5n^3}{mn^7}}$

In the previous example and problem all of the denominators reduced to the point at which no radicals remain. This will not always happen, but as mathematicians we often like to have no radicals in the denominator of a fraction. Clearing any remaining radicals from the denominator of a fraction is called **rationalizing the denominator.** This process uses multiplication on the top and bottom of the fraction to force any radicals in the denominator to reduce completely.

The key to rationalizing the denominator of a fraction is to multiply both the numerator and denominator of the fraction with the right radical expression that will allow the resulting denominator to reduce and be without any remaining radicals. With a single square root this is usually accomplished by multiplying the numerator and denominator by the denominator itself.

EXAMPLE ④ RATIONALIZING THE DENOMINATOR

Simplify the following radical expressions.

a. $\sqrt{\dfrac{3}{5}}$ **b.** $\sqrt{\dfrac{5y}{2x}}$ **c.** $\dfrac{2\sqrt{3n}}{5\sqrt{6m}}$

Solution

a. $\sqrt{\dfrac{3}{5}} = \dfrac{\sqrt{3}}{\sqrt{5}}$ Separate into two radicals.

$\qquad = \dfrac{\sqrt{3}}{\sqrt{5}} \cdot \dfrac{\sqrt{5}}{\sqrt{5}}$ Multiply the numerator and denominator by the denominator. This is the same as multiplying by 1.

$\qquad = \dfrac{\sqrt{15}}{\sqrt{25}}$

$\qquad = \dfrac{\sqrt{15}}{5}$ Reduce the radicals.

b. $\sqrt{\dfrac{5y}{2x}} = \dfrac{\sqrt{5y}}{\sqrt{2x}}$

$\qquad = \dfrac{\sqrt{5y}}{\sqrt{2x}} \cdot \dfrac{\sqrt{2x}}{\sqrt{2x}}$ Multiply the numerator and denominator by the denominator.

$\qquad = \dfrac{\sqrt{10xy}}{\sqrt{4x^2}}$

$\qquad = \dfrac{\sqrt{10xy}}{2x}$

c. $\dfrac{2\sqrt{3n}}{5\sqrt{6m}} = \dfrac{2\sqrt{n}}{5\sqrt{2m}}$

$\quad = \dfrac{2\sqrt{n}}{5\sqrt{2m}} \cdot \dfrac{\sqrt{2m}}{\sqrt{2m}}$ **Multiply the numerator and denominator by the denominator.**

$\quad = \dfrac{2\sqrt{2mn}}{5\sqrt{4m^2}}$

$\quad = \dfrac{2\sqrt{2mn}}{10m}$ **Reduce the radicals.**

$\quad = \dfrac{\sqrt{2mn}}{5m}$ **Reduce the fraction.**

EXAMPLE ④ PRACTICE PROBLEM

Simplify the following radical expressions.

a. $\sqrt{\dfrac{2n}{5m}}$ b. $\dfrac{7x\sqrt{3xy}}{2y\sqrt{5x}}$

    If the denominator contains a higher root, it will take more thought to choose an appropriate expression to multiply by. If you factor the radicand in the denominator, you can then determine what factors are needed to allow the radical to reduce completely. Remember that you want each factor's exponent to be a multiple of the index. If you are working with a cube root, you will want each factor's exponent to be a multiple of 3. This will allow the $\frac{1}{3}$ exponent from the radical to multiply each factor's exponent and get a whole number, thus eliminating the radical from the denominator.

$\dfrac{1}{\sqrt[3]{x}} = \dfrac{1}{\sqrt[3]{x}} \cdot \dfrac{\sqrt[3]{x^2}}{\sqrt[3]{x^2}}$ **Multiplying by $\sqrt[3]{x^2}$ gives you $\sqrt[3]{x^3}$, so the cube root will cancel.**

$\quad = \dfrac{\sqrt[3]{x^2}}{\sqrt[3]{x^3}}$

$\quad = \dfrac{\sqrt[3]{x^2}}{x}$ **The denominator is clear of radicals and thus is rationalized.**

EXAMPLE ⑤ RATIONALIZING DENOMINATORS WITH HIGHER ROOTS

Simplify the following radical expressions.

a. $\sqrt[3]{\dfrac{5xy}{10x^2}}$ b. $\dfrac{2a}{\sqrt[4]{9a^3b^6}}$ c. $\dfrac{\sqrt[3]{2m}}{\sqrt[3]{8np^2}}$

Solution

a. $\sqrt[3]{\dfrac{5xy}{10x^2}} = \sqrt[3]{\dfrac{y}{2x}}$ Reduce the fraction.

$= \dfrac{\sqrt[3]{y}}{\sqrt[3]{2x}}$ Separate the radical.

$= \dfrac{\sqrt[3]{y}}{\sqrt[3]{2x}} \cdot \dfrac{\sqrt[3]{2^2x^2}}{\sqrt[3]{2^2x^2}}$ Multiply by the necessary number of factors to clear the denominator. In this case we need two more 2's and 2 more x's to get the radical to reduce. Note that a cube root will undo a power of 3.

$= \dfrac{\sqrt[3]{4x^2y}}{\sqrt[3]{2^3x^3}}$

$= \dfrac{\sqrt[3]{4x^2y}}{2x}$ Reduce the radicals.

b. $\dfrac{2a}{\sqrt[4]{9a^3b^6}} = \dfrac{2a}{\sqrt[4]{3^2a^3b^6}}$ Factor the radicand.

$= \dfrac{2a}{\sqrt[4]{3^2a^3b^6}} \cdot \dfrac{\sqrt[4]{3^2ab^2}}{\sqrt[4]{3^2ab^2}}$ Multiply by the needed factors. We want two more 3's, one more a, and two more b's. Note that we want eight b's to get an exponent on b that is divisible by 4 (the root's index).

$= \dfrac{2a\sqrt[4]{3^2ab^2}}{\sqrt[4]{3^4a^4b^8}}$

$= \dfrac{2a\sqrt[4]{3^2ab^2}}{3ab^2}$ Reduce the radical.

$= \dfrac{2\sqrt[4]{3^2ab^2}}{3b^2}$ Reduce the fraction.

c. $\dfrac{\sqrt[3]{2m}}{\sqrt[3]{8np^2}} = \dfrac{\sqrt[3]{m}}{\sqrt[3]{4np^2}}$ Reduce the fraction.

$= \dfrac{\sqrt[3]{m}}{\sqrt[3]{2^2np^2}}$ Factor the radicand.

$= \dfrac{\sqrt[3]{m}}{\sqrt[3]{2^2np^2}} \cdot \dfrac{\sqrt[3]{2n^2p}}{\sqrt[3]{2n^2p}}$ Multiply by the needed factors.

$= \dfrac{\sqrt[3]{2mn^2p}}{\sqrt[3]{2^3n^3p^3}}$

$= \dfrac{\sqrt[3]{2mn^2p}}{2np}$ Reduce the radical.

EXAMPLE ⑤ PRACTICE PROBLEM

Simplify the following radical expressions.

a. $\sqrt[3]{\dfrac{24m^2}{15mn^2}}$ **b.** $\dfrac{7x^2}{\sqrt[5]{27xy^7}}$ **c.** $\dfrac{\sqrt[3]{5a^2}}{\sqrt[3]{20ab}}$

CONJUGATES

 CONCEPT INVESTIGATION | WHAT HAPPEND TO THE RADICAL?

Multiply the following expressions together and simplify.

a. $(5 + \sqrt{7})(5 - \sqrt{7})$

b. $(2 - \sqrt{3x})(2 + \sqrt{3x})$

c. $(6 + \sqrt{2a})(6 - \sqrt{2a})$

Describe what happens when you multiply these expressions together.

● ◆

You should be seeing a pattern: When you multiply these expressions together, the result has no radicals remaining. These expressions are special in that they are basically the same except for the sign between the two terms. These expressions are called **conjugates** of one another. Although we did not use the term conjugates, we first saw conjugates in Chapter 4 when we used the quadratic formula. The plus/minus symbol in the quadratic formula resulted in the solutions being conjugates.

> **DEFINITION**
> **Conjugates:**
>
> $$a + b \qquad \text{and} \qquad a - b$$
>
> These expressions are called conjugates of one another. They are simply the sum and difference of the same two terms.
>
> $$5 + x \qquad 5 - x$$
> $$2 + \sqrt{3x} \qquad 2 - \sqrt{3x}$$
> $$-17 + 4\sqrt{5m} \qquad -17 - 4\sqrt{5m}$$

Rationalizing the denominator of a fraction simplifies the fraction so that calculations are easier. Often fractions will have radicals in the denominator with another term. This requires us to use a different approach to rationalizing the denominator. If there are two terms, we will multiply by the conjugate of the denominator to clear all of the radicals in the denominator, which is the goal of rationalizing the denominator.

EXAMPLE   **6** USING CONJUGATES TO RATIONALIZE DENOMINA-TORS

Rationalize the following fractions.

a. $\dfrac{5}{2 + \sqrt{7}}$ **b.** $\dfrac{2 + 3\sqrt{5}}{8 + \sqrt{10}}$ **c.** $\dfrac{2 + \sqrt{x}}{3 - 5\sqrt{x}}$

Solution

a. $\dfrac{5}{2 + \sqrt{7}} = \dfrac{5}{(2 + \sqrt{7})} \cdot \dfrac{(2 - \sqrt{7})}{(2 - \sqrt{7})}$ **Multiply the top and bottom by the conjugate of the denominator.**

$= \dfrac{10 - 5\sqrt{7}}{4 - 7}$

$= \dfrac{10 - 5\sqrt{7}}{-3}$

b. $\dfrac{2 + 3\sqrt{5}}{8 + \sqrt{10}} = \dfrac{(2 + 3\sqrt{5})}{(8 + \sqrt{10})} \cdot \dfrac{(8 - \sqrt{10})}{(8 - \sqrt{10})}$ **Multiply by the conjugate.**

$= \dfrac{16 - 2\sqrt{10} + 24\sqrt{5} - 3\sqrt{50}}{64 - 10}$ **Use the distributive property.**

$= \dfrac{16 - 2\sqrt{10} + 24\sqrt{5} - 15\sqrt{2}}{54}$ **Reduce the radicals.**

c. $\dfrac{2 + \sqrt{x}}{3 - 5\sqrt{x}} = \dfrac{(2 + \sqrt{x})}{(3 - 5\sqrt{x})} \cdot \dfrac{(3 + 5\sqrt{x})}{(3 + 5\sqrt{x})}$ **Multiply by the conjugate.**

$= \dfrac{6 + 10\sqrt{x} + 3\sqrt{x} + 5\sqrt{x^2}}{9 - 25x}$ **Use the distributive property.**

$= \dfrac{6 + 5x + 13\sqrt{x}}{9 - 25x}$ **Reduce the radicals and combine like terms.**

EXAMPLE   **6**   **PRACTICE PROBLEM**

Rationalize the following.

a. $\dfrac{7}{4 + \sqrt{13}}$ **b.** $\dfrac{4 + 5\sqrt{3}}{7 + \sqrt{15}}$ **c.** $\dfrac{3 + \sqrt{a}}{5 - 7\sqrt{ab}}$

Simplifying radical expressions completely requires you to rationalize all denominators and reduce each radical. There should be no fractions remaining inside a radical and no factors left inside a radical that can be pulled out of the radical.

8.3 Exercises

For Exercises 1 through 20, multiply the given radical expressions and simplify the result.

1. $\sqrt{7} \cdot \sqrt{3}$

2. $\sqrt{6} \cdot \sqrt{10}$

3. $\sqrt{5y} \cdot \sqrt{10x}$

4. $\sqrt{8a} \cdot \sqrt{50a}$

5. $\sqrt[3]{4x} \cdot \sqrt[3]{2x^2}$

6. $\sqrt{7m} \cdot \sqrt{14mn}$

7. $\sqrt{3xy} \cdot \sqrt{4y}$

8. $\sqrt{12a^3b} \cdot \sqrt{15a^2b^5}$

9. $\sqrt{18m^5} \cdot \sqrt{2m}$

10. $5\sqrt{3xy} \cdot 7x\sqrt{2y}$

11. $8m\sqrt{14mn} \cdot 2p\sqrt{50m}$

12. $5n^3\sqrt[4]{m^2n^2} \cdot 7m^2n\sqrt[4]{m^3n^7}$

13. $2x^2\sqrt[3]{5x^2y} \cdot 7y\sqrt[3]{4xy}$

14. $2ab\sqrt[3]{a^5b^2} \cdot 7a\sqrt[3]{ab^2}$

15. $12x^2y\sqrt[5]{7x^3y^4} \cdot 3xy^3\sqrt[5]{98x^4y^6}$

16. $8mn\sqrt[5]{2mn^2p} \cdot 5mp\sqrt[5]{16m^3n^3}$

17. $(3 + \sqrt{7})(3 - \sqrt{7})$

18. $(4 - \sqrt{2x})(4 + \sqrt{2x})$

19. $(8 + 5\sqrt{7ab})(8 - 5\sqrt{7ab})$

20. $(2\sqrt{3} + 5\sqrt{7})(2\sqrt{3} - 5\sqrt{7})$

In Exercises 21 through 35, simplify the given radical expressions.

21. $\sqrt{\dfrac{20}{5}}$

22. $\sqrt{\dfrac{432}{75}}$

23. $\sqrt{\dfrac{5x^3}{45xy^2}}$

24. $\sqrt{\dfrac{3a}{4a^3}}$

25. $\sqrt{\dfrac{7cd^2}{3c}}$

26. $\dfrac{5}{\sqrt{3x}}$

27. $\dfrac{5\sqrt{2x}}{3\sqrt{7y}}$

28. $\dfrac{12\sqrt{3ab}}{\sqrt{6a}}$

29. $\sqrt[3]{\dfrac{5x}{7x^2y^2}}$

30. $\sqrt[4]{\dfrac{5ab}{8a^3b^6}}$

31. $\dfrac{5m^2}{\sqrt[3]{2m^4n}}$

32. $\dfrac{\sqrt[3]{4ab^2}}{\sqrt[3]{6a}}$

33. $\dfrac{\sqrt[3]{9x^2y^2}}{7x\sqrt[3]{2xy^3}}$

34. $\dfrac{5ab}{\sqrt[5]{2a^3b^7}}$

35. $\dfrac{8\sqrt[5]{4xy}}{\sqrt[5]{144xy^3z^2}}$

36. $\dfrac{3\sqrt[4]{9xy}}{\sqrt[4]{18x^2y^3}}$

In Exercises 37 through 46, rationalize the denominator of each fraction.

37. $\dfrac{5}{2 + \sqrt{3}}$

38. $\dfrac{7}{8 - \sqrt{6}}$

39. $\dfrac{2 + \sqrt{3}}{5 - \sqrt{7}}$

40. $\dfrac{5}{3 + \sqrt{2x}}$

41. $\dfrac{4 + \sqrt{5}}{7 - 5\sqrt{3}}$

42. $\dfrac{4 + 2\sqrt{7x}}{2 - 3\sqrt{12x}}$

43. $\dfrac{2 + \sqrt{15}}{3 + 4\sqrt{30m}}$

44. $\dfrac{2 + 5\sqrt{m}}{3 - \sqrt{mn}}$

45. $\dfrac{7 - 6\sqrt{ab}}{3 + 5\sqrt{ab}}$

46. $\dfrac{4 + 8\sqrt{5x}}{2 + \sqrt{30x}}$

Solving Radical Equations

So far, we have simplified radical expressions and done arithmetic operations involving radicals. We have also evaluated radicals for different values. Many applications in physics and other sciences use formulas that contain radicals. Often these formulas need to be solved for a variable that is inside the radical itself. We will use the relationship between radicals and rational exponents that we learned in Section 3.1 to help us solve equations involving radicals.

EXAMPLE   ⬛ PERIOD OF A PENDULUM

Many older clocks use pendulums to keep time. The period T (time to go back and forth) of a simple pendulum for a small amplitude is given by the function:

$$T(L) = 2\pi\sqrt{\frac{L}{32}}$$

where T is the period in seconds and L is the length of the pendulum in feet. Use this formula to find the following:

a. The period of a pendulum if its length is 2 ft.

b. How long a pendulum needs to be if you want the period to be 3 seconds.

Solution

a. Because the length of the pendulum is 2 ft, we know that $L = 2$, so we get

$$T(L) = 2\pi\sqrt{\frac{L}{32}}$$

$$T(2) = 2\pi\sqrt{\frac{2}{32}}$$

$$T(2) = 1.57$$

Therefore a 2-ft pendulum will have a period of about 1.57 seconds.

b. We know that $T = 3$, so we get

$$T(L) = 2\pi\sqrt{\frac{L}{32}}$$

$$3 = 2\pi\sqrt{\frac{L}{32}} \qquad \text{Isolate the radical on one side of the equation.}$$

$$\frac{3}{2\pi} = \frac{2\pi\sqrt{\frac{L}{32}}}{2\pi}$$

$$\frac{3}{2\pi} = \sqrt{\frac{L}{32}}$$

$$\left(\frac{3}{2\pi}\right)^2 = \left(\sqrt{\frac{L}{32}}\right)^2 \qquad \text{Square both sides of the equation to eliminate the radical.}$$

In Section 5.2 we discussed solving power equations using reciprocal exponents. We can use this same method to eliminate radicals in most equations. Recall that

$$\sqrt{x} = x^{1/2} \qquad \sqrt[n]{x} = x^{1/n}$$

Using this fact, we can solve radical equations using reciprocal exponents.

SC-Example 1:
Solve the following:

a. $\sqrt{x} = 5$

b. $\sqrt{x + 2} = 3$

c. $5\sqrt[3]{x - 7} = 40$

Solution:

a. $\sqrt{x} = 5$ **Raise both sides to the reciprocal exponent.**

$x^{1/2} = 5$

$(x^{1/2})^2 = 5^2$

$x = 25$

Note that when trying to eliminate a square root, you will always square both sides of the equation.

b. $\sqrt{x + 2} = 3$ **Square both sides.**

$(\sqrt{x + 2})^2 = 3^2$

$x + 2 = 9$ **Solve for x.**

$x = 7$

c. $5\sqrt[3]{x - 7} = 40$ **Isolate the radical by itself.**

$\dfrac{5\sqrt[3]{x - 7}}{5} = \dfrac{40}{5}$

$\sqrt[3]{x - 7} = 8$ **Square both sides.**

$(\sqrt[3]{x - 7})^3 = 8^3$

$x - 7 = 512$ **Solve for x.**

$x = 519$

$$\frac{9}{4\pi^2} = \frac{L}{32}$$

$$32\left(\frac{9}{4\pi^2}\right) = 32\left(\frac{L}{32}\right) \qquad \text{Solve for } L.$$

$$\frac{72}{\pi^2} = L$$

$$7.295 = L$$

We can check this solution with the table.

To have a period of 3 seconds, you will need a pendulum that is about 7.3 feet long.

You should note that this basic formula also works for the period of a child or adult on a swing. You might want to try it.

As you can see, isolating the square root and then squaring both sides will eliminate the radical and allow you to solve for the missing variable.

STEPS TO RADICAL EQUATIONS

1. Isolate a radical on one side of the equation.
2. Square both sides of the equation to eliminate the radical.
3. Isolate any remaining radical if needed and square both sides again.
4. Solve the remaining equation.
5. Check your answer.

(Note: If the radical is not a square root, then raise both sides to the same power as the index of the radical. This will eliminate the radical and allow you to finish solving.)

EXAMPLE 2 THE SPEED OF SOUND

At sea level the speed of sound through air can be calculated by using the following formula.

$$c = 340.3 \sqrt{\frac{T + 273.15}{288.15}}$$

where c is the speed of sound in meters per second and T is the temperature in degrees Celsius. Use this formula to find the following.

a. The speed of sound when the temperature is 30°C.

b. The temperature if the speed of sound is 330 m/s.

Solution

a. $T = 30$, so we get

$$c = 340.3\sqrt{\frac{T + 273.15}{288.15}}$$

$$c = 340.3\sqrt{\frac{30 + 273.15}{288.15}}$$

$$c = 349.05$$

At sea level if the temperature is 30°C, then the speed of sound will be about 349.05 m/s.

b. $c = 330$, so we get

$$c = 340.3\sqrt{\frac{T + 273.15}{288.15}}$$

$$330 = 340.3\sqrt{\frac{T + 273.15}{288.15}}$$

$$\frac{330}{340.3} = \frac{340.3\sqrt{\frac{T + 273.15}{288.15}}}{340.3} \qquad \text{Isolate the square root.}$$

$$0.9697 = \sqrt{\frac{T + 273.15}{288.15}}$$

$$(0.9697)^2 = \left(\sqrt{\frac{T + 273.15}{288.15}}\right)^2 \qquad \begin{array}{l}\text{Square both sides to eliminate}\\\text{the square root.}\end{array}$$

$$0.94038 = \frac{T + 273.15}{288.15}$$

$$288.15(0.94038) = 288.15\left(\frac{T + 273.15}{288.15}\right) \qquad \text{Solve for } T.$$

$$270.9709 = T + 273.15$$

$$-2.179 = T$$

We can check this solution using the graph and trace.

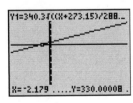

At sea level, if the speed of sound through air is 330 m/s, then the temperature must be about −2.18°C.

EXAMPLE ② PRACTICE PROBLEM

Solve the following equations.

a. $\sqrt{x + 5} = 7$ **b.** $2\sqrt{3x + 4} - 15 = -7$

Some equations may have more than one radical involved and will require more work to solve. You will want to first isolate one of the radicals so that squaring both sides will eliminate that radical and then work on the remaining problem. If you have patience, these problems can be done.

EXAMPLE   3   SOLVING RADICAL EQUATIONS

Solve the following.

a. $\sqrt{x+1} = \sqrt{2x}$

b. $\sqrt{x+10} = 3 + \sqrt{x}$

c. $\sqrt{3x+5} = 2 - \sqrt{3x+2}$

Solution

a.
$$\sqrt{x+1} = \sqrt{2x}$$
$$(\sqrt{x+1})^2 = (\sqrt{2x})^2 \qquad \text{Square both sides to eliminate the radical.}$$
$$x + 1 = 2x$$
$$x + 1 = 2x$$
$$\underline{-x \qquad\qquad -x} \qquad \text{Solve for } x.$$
$$1 = x$$
$$\sqrt{(1)+1} = \sqrt{2(1)} \qquad \text{Check the solution.}$$
$$\sqrt{2} = \sqrt{2}$$

b.
$$\sqrt{x+10} = 3 + \sqrt{x}$$
$$(\sqrt{x+10})^2 = (3 + \sqrt{x})^2 \qquad \textbf{Square both sides to eliminate one of}$$
$$x + 10 = (3 + \sqrt{x})(3 + \sqrt{x}) \qquad \textbf{the radicals.}$$
$$x + 10 = 9 + 3\sqrt{x} + 3\sqrt{x} + x \qquad \textbf{Be sure to use the distributive property.}$$
$$x + 10 = 9 + 6\sqrt{x} + x$$
$$\underline{-x \qquad\qquad\qquad -x} \qquad \text{Isolate the remaining radical.}$$
$$10 = 9 + 6\sqrt{x}$$
$$10 = 9 + 6\sqrt{x}$$
$$\underline{-9 \; -9}$$
$$1 = 6\sqrt{x}$$
$$\frac{1}{6} = \frac{6\sqrt{x}}{6}$$
$$\frac{1}{6} = \sqrt{x}$$
$$\left(\frac{1}{6}\right)^2 = (\sqrt{x})^2 \qquad \textbf{Square both sides again to eliminate the remaining radical.}$$
$$\frac{1}{36} = x$$

X	Y₁	Y₂
.02728	3.1667	3.1667

X=

Check the solution.

c.
$$\sqrt{3x + 5} = 2 - \sqrt{3x + 2}$$

$$(\sqrt{3x + 5})^2 = (2 - \sqrt{3x + 2})^2 \qquad \text{Square both sides.}$$

$$3x + 5 = (2 - \sqrt{3x + 2})(2 - \sqrt{3x + 2}) \qquad \text{Use the distributive property.}$$

$$3x + 5 = 4 - 2\sqrt{3x + 2} - 2\sqrt{3x + 2} + (3x + 2)$$

$$3x + 5 = 4 - 4\sqrt{3x + 2} + 3x + 2 \qquad \text{Simplify.}$$

$$3x + 5 = 6 + 3x - 4\sqrt{3x + 2}$$

$$-1 = -4\sqrt{3x + 2} \qquad \text{Isolate the radical.}$$

$$\frac{1}{4} = \sqrt{3x + 2}$$

$$\left(\frac{1}{4}\right)^2 = (\sqrt{3x + 2})^2 \qquad \text{Square both sides.}$$

$$\frac{1}{16} = 3x + 2$$

$$-\frac{31}{16} = 3x$$

$$-\frac{31}{48} = x$$

X	Y₁	Y₂
-.6458	1.75	1.75

X=

Check the solution.

EXAMPLE ③ PRACTICE PROBLEM

Solve the following.

a. $\sqrt{x + 3} = \sqrt{2x - 8}$ **b.** $\sqrt{2x + 3} = 8 - \sqrt{2x + 1}$

Equations involving radicals can be more complicated and require you to do more solving after eliminating the radicals. After you have eliminated the radicals, use the tools you have learned in other chapters to solve the remaining equation. If a radical other than a square root is found in an equation, remember to use the appropriate reciprocal exponent to eliminate the radical.

EXAMPLE 4 SOLVING RADICAL EQUATIONS

Solve the following.

a. $\sqrt{3x + 1} = 3 + \sqrt{x - 4}$ **b.** $\sqrt{7x + 9} = 5 + \sqrt{2x - 1}$

Solution

a.
$$\sqrt{3x + 1} = 3 + \sqrt{x - 4}$$
$$(\sqrt{3x + 1})^2 = (3 + \sqrt{x - 4})^2$$ Square both sides.
Use the distributive property.
$$3x + 1 = (3 + \sqrt{x - 4})(3 + \sqrt{x - 4})$$
$$1 = 9 + 3\sqrt{x - 4} + 3\sqrt{x - 4} + (x - 4)$$
$$3x + 1 = 5 + x + 6\sqrt{x - 4}$$ Isolate the radical.
$$2x - 4 = 6\sqrt{x - 4}$$
$$(2x - 4)^2 = (6\sqrt{x - 4})^2$$ Square both sides.
Use the distributive property again.
$$(2x - 4)(2x - 4) = 36(x - 4)$$
$$4x^2 - 16x + 16 = 36x - 144$$
$$4x^2 - 52x + 160 = 0$$ Solve the remaining quadratic equation using the quadratic formula.
$$x = 5 \qquad x = 8$$
$$\sqrt{3(5) + 1} = 3 + \sqrt{(5) - 4}$$ Check the solutions.
$$4 = 4$$
$$\sqrt{3(8) + 1} = 3 + \sqrt{(8) - 4}$$
$$5 = 5$$

b.
$$\sqrt{7x + 9} = 5 + \sqrt{2x - 1}$$
$$(\sqrt{7x + 9})^2 = (5 + \sqrt{2x - 1})^2$$ Square both sides.
$$7x + 9 = (5 + \sqrt{2x - 1})(5 + \sqrt{2x - 1})$$ Use the distributive property.
$$7x + 9 = 25 + 5\sqrt{2x - 1} + 5\sqrt{2x - 1} + (2x - 1)$$
$$7x + 9 = 24 + 10\sqrt{2x - 1} + 2x$$ Isolate the radical.
$$5x - 15 = 10\sqrt{2x - 1}$$
$$(5x - 15)^2 = (10\sqrt{2x - 1})^2$$ Square both sides.
$$(5x - 15)(5x - 15) = 100(2x - 1)$$
$$25x^2 - 150x + 225 = 200x - 100$$
$$25x^2 - 350x + 325 = 0$$ Solve the remaining quadratic equation using the quadratic formula.
$$x = 13 \qquad x = 1$$

Only $x = 13$ works. $x = 1$ does not work so it is not a solution to the equation.

Therefore, $x = 13$ is the only solution to this equation.

EXAMPLE ④ PRACTICE PROBLEM

Solve $\sqrt{4x + 1} = 3 + \sqrt{x - 2}$.

E X A M P L E   **5**   SOLVING RADICAL EQUATIONS

Solve the following.

a. $\sqrt{x-5}+7=4$

b. $-5\sqrt{x+4}-10=15$

Solution

a.

$$\sqrt{x-5}+7=4$$

$$\sqrt{x-5}=-3 \qquad \text{Isolate the radical.}$$

$$(\sqrt{x-5})^2=(-3)^2 \qquad \text{Square both sides.}$$

$$x-5=9$$

$$x=14$$

$$\sqrt{(14)-5}+7=4 \qquad \text{Check the solution.}$$

$$10\neq 4 \qquad \text{This solution does not work.}$$

This solution does not work, so there are no solutions to this equation. $x=14$ is an extraneous solution. You should notice that in the second step the square root was equal to a negative number. Because this is not possible, there will be no solution.

b.

$$-5\sqrt{x+4}-10=15$$

$$-5\sqrt{x+4}=25 \qquad \text{Isolate the radical.}$$

$$\sqrt{x+4}=-5$$

$$(\sqrt{x+4})^2=(-5)^2 \qquad \text{Square both sides.}$$

$$x+4=25 \qquad \text{Solve for } x.$$

$$x=21$$

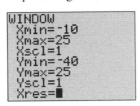

Check the solution.
$x=21$ does not work so it is not a solution.

The solution we found does not work, so we believe there are no solutions to this equation. To check that there are no solutions we can graph both sides of the equation. If the two graphs cross, there would be a solution and we would have to solve the equation again to find our mistake.

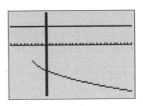

The graphs do not cross, so there are no solutions to this equation.

E X A M P L E   **5**   **P R A C T I C E   P R O B L E M**

Solve $4\sqrt{2x+8}+20=4$.

EXAMPLE ⬛6 SOLVING HIGHER ORDER RADICAL EQUATIONS

Solve the following.

a. $\sqrt[3]{2x} = 5$ **b.** $\sqrt[5]{4x - 9} = 2$

Solution

a.

$$\sqrt[3]{2x} = 5$$

$$\left(\sqrt[3]{2x}\right)^3 = 5^3$$ Because this is a cube root, we will raise both sides to the third power.

$$2x = 125$$

$$\frac{2x}{2} = \frac{125}{2}$$

$$x = 62.5$$

$$\sqrt[3]{2(62.5)} = 5$$

$$5 = 5$$ Check the solution.

b.

$$\sqrt[5]{4x - 9} = 2$$

$$\left(\sqrt[5]{4x - 9}\right)^5 = 2^5$$ Because this is a fifth root, we will raise both sides to the power of 5.

$$4x - 9 = 32$$

$$4x = 41$$

$$x = 10.25$$

$$\sqrt[5]{4(10.25) - 9} = 2$$

$$2 = 2$$

EXAMPLE ⬤6 PRACTICE PROBLEM

Solve $\sqrt[4]{2x + 9} = 3$.

8.4 Exercises

For Exercises 1 through 4, use the following information: Police investigating traffic accidents use the fact that the speed, in miles per hour, of a car traveling on an asphalt road can be determined by the length of the skid marks left by the car after sudden braking. The speed can be modeled by

$$s = \sqrt{30fd}$$

where s is the speed of the car in miles per hour, f is the coefficient of friction for the road, and d is the length in feet of the skid marks.

1. Find the speed of a car if it leaves skid marks 150 ft long on a dry asphalt road that has a coefficient of friction equal to 1.

2. Find the speed of a car if it leaves skid marks 120 ft long on an asphalt road with a coefficient of friction equal to 0.444.

3. Find the coefficient of friction of a road if a car traveling 55 mph leaves skid marks 115 ft long after sudden braking. (Round to three decimal places.)

4. Find the length of skid marks a police officer would expect to find if a person traveling 25 mph suddenly brakes on a road with a coefficient of friction equal to 1.

5. Find the period of a pendulum if its length is 3 ft. See Example 1.

6. Find the period of a pendulum if its length is 10 ft.

7. If you want a clock's pendulum to have a period of 1 second, what must its length be?

8. If you want a clock's pendulum to have a period of 2 seconds, what must its length be?

For Exercises 9 through 12, use the following information: When an object is dropped from a height of h feet, the time it will take to hit the ground can be approximated by the formula

$$t = \sqrt{\frac{2h}{32}}$$

where t is the time in seconds for the object to fall h feet.

9. Find how long it will take an object to fall 100 feet.

10. Find how long it will take an object to fall 50 feet.

11. If you want an object to fall for 10 seconds, from what height should the object be dropped?

12. If you want an object to fall for 20 seconds, from what height should the object be dropped?

13. Find the speed of sound at sea level when the temperature is 28°C. See Example 2.

14. Find the temperature at sea level if the speed of sound is 345 m/s.

15. In Exercise 1 of Section 8.1 we used the function

$$C(h) = 2.29\sqrt{h} + 3.2$$

to calculate the monthly cost in millions of dollars to produce h hundred thousand space heaters at Jim Bob's Heater Source.

a. How many heaters can Jim Bob's make in a month if they have a budget of $15 million?

b. How many heaters can they make if they have a budget cut and can spend only $12 million?

16. In Exercise 3 of Section 8.1 we used the function

$$L(M) = 0.0165 \sqrt[3]{M}$$

to determine the leg length in meters of a basilisk lizard with a body mass of M grams.

a. Use this model to estimate the live body mass of a museum specimen that has a leg length of 0.1 meters.

b. Use the model to estimate the live body mass of a specimen that has a leg length of 0.075 meters.

17. In Exercise 4 of Section 8.1 we used the function

$$L(M) = 0.330 \sqrt[3]{M}$$

where $L(M)$ represents the body length in meters of a mammal with a body mass of M kilograms.

a. Use this model to estimate the body mass of a mammal whose body length is 1 m.

b. Use this model to estimate the body mass of a mammal whose body length is 2.4 m.

18. In Exercise 5 of Section 8.1 we were given the following formula for the relationship between the thoracic mass of the butterfly and its mean forward airspeed.

$$V(m) = 17.6 \sqrt[5]{m^3}$$

where $V(m)$ represents the mean forward airspeed in meters per second of a butterfly with a thoracic mass of m grams.

a. Use the model to estimate the thoracic mass of a butterfly that can fly at 5 meters per second.

b. Use the model to estimate the thoracic mass of a butterfly that can fly at 6.2 meters per second.

For Exercises 19 through 58, solve each equation. Check your solutions with a table or graph.

19. $\sqrt{x + 2} = 5$

20. $\sqrt{w - 7} = 10$

21. $\sqrt{3x + 4} = 15$

22. $\sqrt{7x + 1} = 13$

23. $\sqrt{-2t + 4} = 7$

24. $\sqrt{-4m - 9} = 6$

25. $3\sqrt{2x + 5} + 12 = 24$

26. $-2\sqrt{3x - 7} - 11 = -19$

27. $5\sqrt{2x} = 40$

28. $-0.25\sqrt{2.3w} = -14$

29. $2.4\sqrt{2.5g + 4} - 7.5 = 4.6$

30. $3.5\sqrt{1.7x + 6} + 2.8 = 14.3$

31. $\sqrt{2x + 5} = \sqrt{3x - 4}$

32. $\sqrt{p + 6} = \sqrt{3p + 2}$

33. $\sqrt{4 + 2z} = \sqrt{3 - 7z}$

34. $\sqrt{2 + 5x} = \sqrt{3x + 10}$

35. $\sqrt{x - 7} - \sqrt{x} = -1$

36. $\sqrt{x - 11} - \sqrt{x} = -1$

37. $\sqrt{2x + 3} = 1 - \sqrt{x + 5}$

38. $\sqrt{6x - 5} = 4 + \sqrt{5x + 2}$

39. $\sqrt{4x^2 + 9x + 5} = 5 + x$

40. $\sqrt{12x^2 + 5x - 8} = 7 + x$

41. $\sqrt{w + 2} = 4 - w$

42. $\sqrt{-m + 2} = m - 2$

43. $\sqrt{6x + 1} - \sqrt{9x} = -1$

44. $\sqrt{2x + 11} + \sqrt{2x + 7} + 4 = 0$

45. $\sqrt{6x + 5} + \sqrt{3x + 2} = 5$

46. $\sqrt{3x - 5} - \sqrt{3x + 3} = 4$

47. $\sqrt{8x + 8} - \sqrt{8x - 4} = -2$

48. $\sqrt{5x + 4} - \sqrt{5x - 7} = -9$

49. $\sqrt{2x + 3} = 1 + \sqrt{x + 1}$

50. $\sqrt{2x + 5} - 3 = \sqrt{x - 2}$

51. $\sqrt[3]{2x + 5} = 6$

52. $\sqrt[5]{x - 8} = 4$

53. $\sqrt[4]{5x - 9} = 2$

54. $\sqrt[7]{145x} = 6$

55. $\sqrt[3]{5x + 2} + 7 = 12$

56. $\sqrt[5]{2x - 15} + 9 = 4$

57. $\sqrt[3]{-4x + 3} = -5$

58. $\sqrt[5]{-2x + 20} = -4$

SECTION

8.5 Complex Numbers

DEFINITION OF IMAGINARY AND COMPLEX NUMBERS

Throughout our work with radicals we have noted that the square root of a negative number is not a real number. We saw this in the chapter on quadratics when we found no real solutions to some problems. $\sqrt{-4}$ is considered a nonreal number and is called an **imaginary number.** Imaginary numbers got their name because when they were first discovered, many mathematicians did not believe that they existed. Later they were proven to exist, and they have been shown to be applicable in different fields of mathematics and science. Although imaginary numbers were proven to exist, their name had stuck by then, so we must remember that the name *imaginary* does not mean that these numbers do not exist. In mathematics the number $\sqrt{-1}$ is the **imaginary unit** and is usually represented by the letter i. Using the letter i, we can represent imaginary numbers without showing a negative under a square root.

EXAMPLE **1** SIMPLIFYING RADICALS INTO COMPLEX NUMBERS

Simplify the following, using the imaginary number i.

a. $\sqrt{-25}$

b. $\sqrt{-100}$

c. $\sqrt{-30}$

d. $-\sqrt{16}$

Solution

a. $\sqrt{-25} = \sqrt{-1 \cdot 25}$

$= \sqrt{-1} \cdot \sqrt{25}$ Factor out the negative 1.

$= i \cdot 5$

$= 5i$

b. $10i$

c. $i\sqrt{30} \approx 5.477i$

d. $-\sqrt{16} = -4$. This is not an imaginary number because the negative is not inside the square root.

DEFINITION

Imaginary unit:

$$\sqrt{-1} = i$$

Imaginary numbers can be combined with the real numbers into what are called **complex numbers.** Any number that can be written in the form $a + bi$ is a complex number; a is considered the real part of a complex number, and b is considered the imaginary part. All real numbers are considered complex numbers who's imaginary part is equal to zero. All imaginary numbers are also considered complex numbers who's real part is equal to zero. Complex numbers are used in many areas of mathematics and many physics and engineering fields. The shapes of airplane wings are developed and studied by using complex numbers. Many areas of algebra, such as fractals and chaos theory, can be studied by working in the complex number system.

DEFINITION

Complex number: Any number that can be written in the form

$$a + bi$$

where a and b are real numbers.

EXAMPLE **2** NAME THE PARTS OF COMPLEX NUMBERS

For each complex number, name the real part and the imaginary part.

a. $5 + 4i$ **b.** $-3 + 7i$

c. $5i$ **d.** 9

Solution

a. Real part $= 5$; imaginary part $= 4$.

b. Real part $= -3$; imaginary part $= 7$.

c. Real part $= 0$; imaginary part $= 5$.

d. Real part $= 9$; imaginary part $= 0$.

OPERATIONS WITH COMPLEX NUMBERS

Complex numbers can be added and subtracted easily by adding or subtracting the real parts together and then adding or subtracting the imaginary parts together. This is similar to combining like terms with variables.

ADDING COMPLEX NUMBERS

$$(a + bi) + (c + di)$$
$$(a + c) + (b + d)i$$

When adding complex numbers, add the real parts together and then add the imaginary parts.

SUBTRACTING COMPLEX NUMBERS

$$(a + bi) - (c + di)$$
$$(a - c) + (b - d)i$$

When subtracting complex numbers, subtract the real parts and then subtract the imaginary parts. Be careful with the positive and negative signs.

EXAMPLE 3 ADD OR SUBTRACT COMPLEX NUMBERS

Add or subtract the following complex numbers.

a. $(2 + 8i) + (6 + 7i)$ **b.** $(5 - 4i) + (7 + 6i)$

c. $(6 + 3i) - (4 + 8i)$ **d.** $(2.5 + 3.8i) - (4.6 - 7.2i)$

Solution

a. $(2 + 8i) + (6 + 7i) = 2 + 6 + 8i + 7i$ Add the real parts
$$= (2 + 6) + (8i + 7i)$$ Add the imaginary parts.
$$= 8 + 15i$$

b. $12 + 2i$

c. $(6 + 3i) - (4 + 8i) = 6 + 3i - 4 - 8i$ Distribute the negative.

Subtract the real parts.
$$= (6 - 4) + (3i - 8i)$$ Subtract the imaginary parts.
$$= 2 - 5i$$

Be sure to watch your negative signs.

d. $-2.1 + 11i$

EXAMPLE 3   PRACTICE PROBLEM

Add or subtract the following complex numbers.

a. $(2 + 6i) + (-3 + 5i)$ **b.** $(3.4 - 7.5i) - (-4.2 - 5.9i)$

c. $(-7 + 6i) + (-3 - 4i)$ **d.** $(-3 + 5i) - (4 + 9i)$.

Because $i = \sqrt{-1}$, we can calculate other powers of i by considering the following pattern.

$$i = \sqrt{-1}$$
$$i^2 = (\sqrt{-1})^2 = -1$$
$$i^3 = -1(\sqrt{-1}) = -\sqrt{-1} = -i$$
$$i^4 = i^2 i^2 = (-1)(-1) = 1$$
$$i^5 = 1i = i$$

This pattern will continue to repeat as you have higher and higher powers of i. If you note that $i^4 = 1$, then you can basically reduce any powers of i by a multiple of 4 without changing the resulting value. For example,

$$i^5 = i^{5-4} = i \qquad \text{Reduce the exponent by 4.}$$
$$i^{10} = i^{10-8} = i^2 = -1 \quad \text{Reduce the exponent by 8 and simplify.}$$

In most problems we will deal with, the most important power of i that you should know and use is $i^2 = -1$. When you multiply complex numbers, using this fact will help you to reduce the answers to complex form. Whenever you see i^2 in a calculation, you should replace it with a -1 and continue to combine like terms and simplify.

MULTIPLYING COMPLEX NUMBERS

$$(a + bi)(c + di)$$

When multiplying complex numbers, use the distributive property, and change any i^2 to -1. Simplify the result by combining like terms and writing the result in the standard form of a complex number.

EXAMPLE   **4** MULTIPLYING COMPLEX NUMBERS

Multiply the following complex numbers.

a. $3(4 + 9i)$ **b.** $2i(7 - 3i)$

c. $(2 + 5i)(4 + 8i)$ **d.** $(3 + 2i)(5 - 6i)$

Solution

In all of these problems you will use the distributive property and simplify where possible.

a. $12 + 27i$ Distribute the 3.

b. $2i(7 - 3i)$ Distribute.

   $14i - 6i^2$

   $14i - 6(-1)$ Replace i^2 with -1.

   $14i + 6$ Simplify.

   $6 + 14i$ Put in standard complex form $a + bi$.

c. $(2 + 5i)(4 + 8i)$

   $8 + 16i + 20i + 40i^2$ Use the distributive property.

   $8 + 36i + 40i^2$ Combine like terms.

   $8 + 36i + 40(-1)$ Replace i^2 with -1.

   $8 + 36i - 40$ Simplify.

   $-32 + 36i$

d. $(3 + 2i)(5 - 6i)$

   $15 - 18i + 10i - 12i^2$ Use the distributive property.

$$15 - 8i - 12i^2 \qquad \text{Combine like terms.}$$
$$15 - 8i - 12(-1) \qquad \text{Replace } i^2 \text{ with } -1.$$
$$15 - 8i + 12 \qquad \text{Simplify.}$$
$$27 - 8i$$

EXAMPLE ④ PRACTICE PROBLEM

Multiply the following complex numbers.

a. $3i(4 + 8i)$ **b.** $(2 + 7i)(3 + 5i)$

c. $(4 - 2i)(7 + 9i)$ **d.** $(2 + 5i)(2 - 5i)$

Example 4 Practice Problem part d is an example of multiplying two complex numbers that have a very special relationship. These two complex numbers are what are called **complex conjugates** of one another. When complex conjugates are multiplied together, you should note that the product is a real number and therefore has no imaginary part remaining.

DEFINITION

Complex conjugates:

$$a + bi \quad \text{and} \quad a - bi$$

are conjugates of one another. The conjugate of a complex number is that complex number with the sign of the imaginary number changed.

$$5 - 3i \quad \text{and} \quad 5 + 3i$$

are complex conjugates of one another.

EXAMPLE ⑤ WRITING COMPLEX CONJUGATES

Write the complex conjugate of the following.

a. $2 + 9i$ **b.** $4 - 6i$

c. $3.4i$ **d.** 10

Solution

a. $2 - 9i$

b. $4 + 6i$

c. $-3.4i$

d. 10 Because the imaginary part is zero, there is no other complex conjugate.

EXAMPLE **6** MULTIPLYING BY COMPLEX CONJUGATES

Multiply the following complex numbers by their conjugates.

a. $2 + 9i$ **b.** $4 - 6i$ **c.** $3.4i$

Solution

a. $(2 + 9i)(2 - 9i)$

$4 - 18i + 18i - 81i^2$

$4 - 81i^2$

$4 - 81(-1)$

$4 + 81$

85

b. $(4 - 6i)(4 + 6i)$

$16 + 24i - 24i - 36i^2$

$16 - 36i^2$

$16 - 36(-1)$

$16 + 36$

52

c. $3.4i(-3.4i)$

$-11.56i^2$

$-11.56(-1)$

11.56

As you can see from Example 6, when complex conjugates are multiplied together, the imaginary parts cancel out, and you are left with a real number. This is a very helpful attribute of conjugates when complex numbers are involved in a division problem. In dealing with division or fractions, it is standard practice to only allow a denominator with real numbers. Imaginary numbers need to be eliminated from any denominator. This can be done by multiplying both the numerator and the denominator by the conjugate of the denominator (similar to what you did to rationalize denominators). This will not change the value of the fraction but will change the denominator into a real number. The resulting fraction can then be written in standard complex form..

DIVIDING COMPLEX NUMBERS

$$\frac{a + bi}{c + di}$$

$$\frac{(a + bi)}{(c + di)} \cdot \frac{(c - di)}{(c - di)}$$

In dividing complex numbers, you must clear all denominators of imaginary numbers. This is done by multiplying both the denominator and the numerator by the conjugate of the denominator. Remember to reduce the final answer and write it in the standard form for complex numbers.

EXAMPLE   7   DIVIDING COMPLEX NUMBERS

Divide the following. Put all answers in standard complex form.

a. $\dfrac{10 + 8i}{2}$

b. $\dfrac{4 + 7i}{5 + 3i}$

c. $\dfrac{2 - 9i}{4 + 3i}$

d. $\dfrac{6 - 7i}{5i}$

Solution

a. $\dfrac{10 + 8i}{2} = \dfrac{10}{2} + \dfrac{8}{2}i$ Reduce the fraction and put into the standard form of a complex number.

$= 5 + 4i$ Simplify.

b. $\dfrac{4 + 7i}{5 + 3i}$

$\dfrac{(4 + 7i)}{(5 + 3i)} \cdot \dfrac{(5 - 3i)}{(5 - 3i)}$ Multiply the numerator and denominator by the conjugate of the denominator.

$\dfrac{20 - 12i + 35i - 21i^2}{25 - 15i + 15i - 9i^2}$

$\dfrac{20 + 23i - 21(-1)}{25 - 9(-1)}$

$\dfrac{20 + 23i + 21}{25 + 9}$

$\dfrac{41 + 23i}{34}$

$\dfrac{41}{34} + \dfrac{23}{34}i$ Write in the standard form of a complex number.

c. $\dfrac{2 - 9i}{4 + 3i}$

$\dfrac{(2 - 9i)}{(4 + 3i)} \cdot \dfrac{(4 - 3i)}{(4 - 3i)}$ Multiply the numerator and denominator by the conjugate of the denominator.

$\dfrac{8 - 6i - 36i + 27^2}{16 - 12i + 12i - 9i^2}$

$\dfrac{8 - 42i - 27}{16 + 9}$

$\dfrac{-19 - 42i}{25}$

$-\dfrac{19}{25} - \dfrac{42}{25}i$ Write in the standard form of a complex number.

d. $\dfrac{6 - 7i}{5i}$

$\dfrac{(6 - 7i)}{5i} \cdot \dfrac{i}{i}$ Multiply the numerator and denominator by i. Because the denominator does not have a real part we only need to multiply by i to rationalize the denominator.

$\dfrac{6i - 7i^2}{5i^2}$

$\dfrac{6i + 7}{-5}$

$-\dfrac{7}{5} - \dfrac{6}{5}i$ Write in the standard form of a complex number.

EXAMPLE ⑦ PRACTICE PROBLEM

Divide the following. Put all answers in standard complex form.

a. $\dfrac{14 + 8i}{2}$ **b.** $\dfrac{4 - 7i}{3i}$

c. $\dfrac{4 + 2i}{3 - 7i}$ **d.** $\dfrac{7.2 + 3.4i}{1.4 + 5.6i}$

SOLVING EQUATIONS WITH COMPLEX SOLUTIONS

Many equations will have complex solutions. The most common place in which we will see these types of solutions is in working with quadratics. The quadratic formula is a great tool to find both real and complex solutions to any quadratic equation. Now when a discriminant is a negative number, we can write our solutions using complex numbers instead of just saying that there are no real solutions. This results in a more complete answer to the equation. You will notice that if a complex number is a solution to a polynomial, the complex conjugate will also be a solution to that equation. In most applications we are interested only in the real solutions to an equation, but in some contexts, such as electrical engineering, complex solutions are of interest.

EXAMPLE 8 SOLVING EQUATIONS WITH COMPLEX SOLUTIONS

Solve the following equations. Give answers in the standard form of a complex number.

a. $t^2 + 2t + 5 = 0$ **b.** $x^2 = -25$

c. $x^2 + 4x = -30$ **d.** $x^3 - 10x^2 + 29x = 0$

Solution

a. $t^2 + 2t + 5 = 0$

$t = \dfrac{-2 \pm \sqrt{2^2 - 4(1)(5)}}{2(1)}$ Use the quadratic formula.

$$t = \frac{-2 \pm \sqrt{-16}}{2}$$

The discriminant is -16, so the answer will be a complex number.

$$t = \frac{-2 \pm 4i}{2}$$

$$t = -1 + 2i \qquad t = -1 - 2i$$

Write in standard form.

b. $\qquad x^2 = -25$

$$x = \pm\sqrt{-25}$$

Use the square root property.

$$x = \pm 5i$$

c. $\qquad x^2 + 4x = -30$

$$x^2 + 4x + 30 = 0$$

$$x = \frac{-4 \pm \sqrt{4^2 - 4(1)(30)}}{2(1)}$$

$$x = \frac{-4 \pm \sqrt{-104}}{2}$$

$$x = \frac{-4 \pm 2i\sqrt{26}}{2}$$

$$x = -2 + i\sqrt{26} \qquad x = -2 - i\sqrt{26}$$

$$x \approx -2 + 5.1i \qquad x \approx -2 - 5.1i$$

d. $\qquad x^3 - 10x^2 + 29x = 0$

Factor the common term out.

$$x(x^2 - 10x + 29) = 0$$

Set each factor equal to zero and continue to solve.

$$x = 0 \qquad x^2 - 10x + 29 = 0$$

$$x = 0 \qquad x = \frac{-(-10) \pm \sqrt{(-10)^2 - 4(1)(29)}}{2(1)}$$

$$x = 0 \qquad x = \frac{10 \pm \sqrt{100 - 116}}{2}$$

$$x = 0 \qquad x = \frac{10 \pm \sqrt{-16}}{2}$$

$$x = 0 \qquad x = \frac{10 \pm 4i}{2}$$

$$x = 0 \qquad x = 5 + 2i \qquad x = 5 - 2i$$

EXAMPLE ⑧ PRACTICE PROBLEM

Solve the following equations. Give answers in the standard form of a complex number.

a. $w^2 + 6w + 18 = 0$ 　　　　**b.** $x^2 = -81$

c. $x^2 + 8x = -18$ 　　　　**d.** $x^3 - 4x^2 + 13x = 0$

8.5 Exercises

In Exercises 1 through 10, simplify the given expression and write your answer in terms of i.

1. $\sqrt{-100}$

2. $5 + \sqrt{-36}$

3. $\sqrt{-49} + \sqrt{-25}$

4. $\sqrt{9} + \sqrt{-16}$

5. $\sqrt{4} - \sqrt{-16}$

6. $\sqrt{-200}$

7. $\sqrt{-3} + \sqrt{-5.6}$

8. $\sqrt{2}(5 + \sqrt{-8})$

9. $\sqrt{7}(\sqrt{28} + \sqrt{-63})$

10. $(2 + \sqrt{5})(3 - \sqrt{-7})$

In Exercises 11 through 47, simplify the given expression and write your answer in the standard form for a complex number.

11. $(2 + 5i) + (6 + 4i)$

12. $(3 - 8i) + (2 + 6i)$

13. $(5.6 + 3.2i) + (2.3 - 4.9i)$

14. $(3 + 2i) - (5 + 8i)$

15. $(5i) + (3 - 9i)$

16. $(2 - 7i) - (5 - 6i)$

17. $(5 + 7i) - 10$

18. $(3 - 8i) - (10 - 12i)$

19. $(3 + 5i) - (7 + 8i)$

20. $(-4 - 6i) - (3 - 9i)$

21. $(4.7 - 3.5i) + (1.8 - 5.7i)$

22. $(2.4 - 9.6i) - (3.5 + 8.6i)$

23. $(2 + 8i) + (3 - 8i)$

24. $(4 - 7i) - (4 - 7i)$

25. $(2 + 3i)(4 - 7i)$

26. $(2 - 4i)(3 - 5i)$

27. $(2.4 - 3.5i)(4.1 - 2.6i)$

28. $(5 + 4i)(3 + 2i)$

29. $(2.4 + 8.1i)(1.3 + 4.5i)$

30. $(5 + 2i)(5 + 2i)$

31. $(3 - 7i)(3 + 7i)$

32. $(2.3 + 4.5i)(2.3 - 4.5i)$

33. $(2 + 9i)(2 + 9i)$

34. $(3.3 - 4.4i)(3.3 - 4.4i)$

35. $\dfrac{12 + 9i}{3}$

36. $\dfrac{45 - 15i}{5}$

37. $\dfrac{4 + 7i}{2i}$

38. $\dfrac{5 + 8i}{3i}$

39. $\dfrac{2.4 + 3.78i}{7.2i}$

40. $\dfrac{3.45 + 8.29i}{2.47i}$

41. $\dfrac{5 + 2i}{2 + 3i}$

42. $\dfrac{4 + 6i}{2 - 7i}$

43. $\dfrac{2.3 - 4.1i}{3.4 - 7.3i}$

44. $\dfrac{3.5 + 7.2i}{2.3 + 4.1i}$

45. $\dfrac{-8.3 - 4.5i}{-2 + 4.7i}$

46. $\dfrac{-7.2 - 8.6i}{-3.5 - 5.73i}$

47. $\dfrac{-12 - 73i}{-5 - 14i}$

48. $\dfrac{-8 - 14i}{-11 - 9i}$

In Exercises 49 through 62, solve each equation and write any complex solutions in standard form.

49. $x^2 + 9 = 0$

50. $w^2 + 25 = 0$

51. $3t^2 + 45 = -3$

52. $2x^2 - 18 = 0$

53. $3(x - 8)^2 + 12 = 4$

54. $-0.25(t + 3.4)^2 + 9 = 15$

55. $2x^2 + 5x + 4 = -4$

56. $3x^2 + 2x + 7 = -8$

57. $5x^3 + 2x^2 - 7x = 0$

58. $8x^3 + 4x^2 - 14x = 0$

59. $0.25x^3 - 3.5x^2 + 16.25x = 0$

60. $3x^3 + 24x^2 + 53x = 0$

61. $2x^2 - 28x + 45 = -54$

62. $-3w^2 + 12w - 26 = 0$

Chapter 8 Summary

Section 8.1 Introduction to Radical Functions

- A **radical** is square root $\sqrt{x}$ or higher root $\sqrt[n]{x}$.
- **Radical functions** can be used to model some real-life situations.
- **Higher roots** can be calculated on the calculator by using rational exponents. Use parentheses around all rational exponents.
- When not in a context, the domain of an even root is restricted to values that keep the radicand nonnegative.
- When not in a context, the domain and range of odd roots are typically all real numbers.

EXAMPLE 1

The distance to the horizon at a particular altitude can be modeled by the function $H(a) = \sqrt{1.5a}$, where $H(a)$ is the distance to the horizon in miles at an altitude a feet above sea level. Use this equation to determine the distance to the horizon if you are standing on top of the Empire State Building at 1250 feet.

© Bill Ross/CORBIS

Solution

$H(1250) = 43.3$. The horizon will be 43.3 miles away when you are on the top of the Empire State Building.

EXAMPLE 2

Give the domain and range of the function.

a. $f(x) = -\sqrt{x + 6}$

b. $g(x) = \sqrt[5]{3x}$

Solution

a. For the domain the radicand of an even root must be nonnegative, so

$$x + 6 \geq 0$$
$$x \geq -6$$

We confirm this with the graph, and the range is also shown.

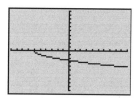

Domain: $[-6, \infty)$
Range: $(-\infty, 0]$

b. Domain: All real numbers
Range: All real numbers

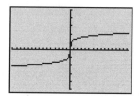

Section 8.2 Simplifying, Adding, and Subtracting Radicals

- The **principal root** is the positive root of a square root.
- To get a **negative root,** we put a negative sign outside the square root.
- **Negative radicands** result in nonreal solutions when in radicals with an even index.
- **Simplifying radicals** can be done by using the rules for exponents or by factoring into powers equal to the index of the radical.
- To **add or subtract radicals,** they need to have the same radicand and index.
- Simplify radicals before you add or subtract. The radicands do not change in adding or subtracting a radical only the coefficients will be added or subtracted.

EXAMPLE 3

Simplify the following radicals.

a. $\sqrt{12x^3y^8}$ **b.** $\sqrt[5]{32a^4b^{10}c^7}$

Solution

a.
$$\sqrt{12x^3y^8}$$
$$\sqrt{3}\cdot\sqrt{2^2}\cdot\sqrt{x^2}\cdot\sqrt{x}\cdot\sqrt{y^2}\cdot\sqrt{y^2}\cdot\sqrt{y^2}\cdot\sqrt{y^2}$$
$$2xyyyy\sqrt{3x}$$
$$2xy^4\sqrt{3x}$$

b. $\sqrt[5]{32a^4b^{10}c^7}$

$$(2^5a^4b^{10}c^7)^{\frac{1}{5}}$$

$$2a^{\frac{4}{5}}b^2c^{\frac{7}{5}}$$

$$2a^{\frac{4}{5}}b^2c^{1\frac{2}{5}}$$

$$2b^2c\sqrt[5]{a^4c^2}$$

EXAMPLE 4

Perform the indicated operation and simplify.

a. $5\sqrt{x}+3\sqrt{x}$

b. $\sqrt{8x^3y}-7x\sqrt{2xy}$

Solution

a. $5\sqrt{x}+3\sqrt{x}=8\sqrt{x}$

b. $\sqrt{8x^3y}-7x\sqrt{2xy}$

$$2x\sqrt{2xy}-7x\sqrt{2xy}$$

$$-5x\sqrt{2xy}$$

Section 8.3 Multiplying and Dividing Radicals

- When **multiplying radicals** with the same index, you multiply the radicands together, and anything outside each radical is also multiplied together.
- **Conjugates** are expressions that are the sum and difference of the same two terms: $(a+b)$ and $(a-b)$.
- Fractions inside radicals can either be reduced inside the radical or separated into distinct radicals and then reduced.
- Fractions that have a radical in the denominator are usually rationalized.

- To **rationalize** a denominator of a fraction, use one of these techniques:

  1. When a single square root is in the denominator, you simply multiply both the numerator and the denominator by the original radical in the denominator.

  2. When higher roots are in the denominator, you multiply by the root with the same index with the needed factors to make the radical reduce completely.

3. If two terms with a radical are in the denominator, you will need to use conjugates to eliminate the radical.

EXAMPLE 5

Perform the indicated operations and simplify. Rationalize all denominators.

a. $2w\sqrt{3r} \cdot 5\sqrt{7wr}$ **b.** $\sqrt{\dfrac{4}{7x}}$

c. $\dfrac{7}{\sqrt[3]{5x^2}}$ **d.** $\dfrac{2 + \sqrt{3}}{5 - \sqrt{7}}$

Solution

a. $2w\sqrt{3r} \cdot 5\sqrt{7wr}$

$10w\sqrt{21wr^2}$

$10wr\sqrt{21w}$

b. $\sqrt{\dfrac{4}{7x}} \cdot \dfrac{\sqrt{7x}}{\sqrt{7x}}$

$\dfrac{\sqrt{28x}}{7x}$

c. $\dfrac{7}{\sqrt[3]{5x^2}} \cdot \dfrac{\sqrt[3]{25x}}{\sqrt[3]{25x}}$

$\dfrac{7\sqrt[3]{25x}}{5x}$

d. $\dfrac{2 + \sqrt{3}}{5 - \sqrt{7}} \cdot \dfrac{5 + \sqrt{7}}{5 + \sqrt{7}}$

$\dfrac{10 + 2\sqrt{7} + 5\sqrt{3} + \sqrt{21}}{25 + 5\sqrt{7} - 5\sqrt{7} - 7}$

$\dfrac{10 + 2\sqrt{7} + 5\sqrt{3} + \sqrt{21}}{18}$

Section 8.4 Solving Radical Equations

- When **solving an equation with a radical,** isolate the radical on one side of the equation and then square both sides of the equation.

- If there is more than one radical, you might need to square both sides of the equation twice. Remember to always isolate a radical before you square both sides.

- If a radical is not a square root, then raise both sides of the equation to the same power as the index of the radical.

EXAMPLE 6

The distance to the horizon at a particular altitude can be modeled by the function $H(a) = \sqrt{1.5a}$, where $H(a)$ is the distance to the horizon in miles at an altitude a feet above sea level. Use this function to find the altitude you must be at for the horizon to be 100 miles away.

Solution

$$100 = \sqrt{1.5a}$$
$$100^2 = (\sqrt{1.5a})^2$$
$$10000 = 1.5a$$
$$6666.67 = a$$

For the horizon to be 100 miles away, you must be about 6666.67 feet above sea level.

EXAMPLE 7

Solve the following equations.

a. $\sqrt{5x - 6} = 7 + \sqrt{x}$

b. $\sqrt[4]{2x + 8} + 9 = 20$

Solution

a.
$$\sqrt{5x - 6} = 7 + \sqrt{x}$$
$$(\sqrt{5x - 6})^2 = (7 + \sqrt{x})^2$$
$$5x - 6 = (7 + \sqrt{x})(7 + \sqrt{x})$$
$$5x - 6 = 49 + 7\sqrt{x} + 7\sqrt{x} + x$$
$$5x - 6 = 49 + 14\sqrt{x} + x$$
$$4x - 55 = 14\sqrt{x}$$
$$(4x - 55)^2 = (14\sqrt{x})^2$$
$$16x^2 - 440x + 3025 = 196x$$
$$16x^2 - 636x + 3025 = 0$$

$x = 34.23$ $x = 5.52$ **Only one of these answers is valid.**

$x = 34.23$

Check your answers.

X	Y₁	Y₂
34.23	12.851	12.851
5.52	4.6476	9.3495

X=

$(\sqrt[4]{2x + 8})^4 = (11)^4$

$2x + 8 = 14641$

$x = 7316.5$

X	Y₁	Y₂
7316.5	20	20

X=

b. $\sqrt[4]{2x + 8} + 9 = 20$

$\sqrt[4]{2x + 8} = 11$

Section 8.5 Complex Numbers

- The imaginary unit is $i = \sqrt{-1}$.
- When a negative number is the radicand of a square root, it results in an imaginary number.
- A standard complex number is $a + bi$, where a is the real part and b is the imaginary part.
- Complex conjugates are $a + bi$ and $a - bi$.
- To add or subtract complex numbers, combine the real parts and combine the imaginary parts.
- To multiply complex numbers, use the distributive property if necessary, and use the fact that $i^2 = -1$ to reduce any higher powers of the imaginary unit.
- When dividing complex numbers, you can rationalize the denominator by multiplying both the numerator and denominator by the complex conjugate.
- When solving a quadratic equation, the quadratic formula can be used to find any complex solutions.

EXAMPLE 8

Perform the indicated operation and answer in standard complex form.

a. $(2 + 5i) + (7 - 8i)$ **b.** $(3 - 7i)(4 + 2i)$

c. $\dfrac{5 + 3i}{2i}$ **d.** $\dfrac{4 - 3i}{2 + 5i}$

Solution

a. $(2 + 5i) + (7 - 8i) = 9 - 3i$

b. $(3 - 7i)(4 + 2i)$

$12 + 3i - 28i - 14i^2$

$12 - 25i + 14$

$26 - 25i$

c. $\dfrac{5 + 3i}{2i} \cdot \dfrac{i}{i}$

$\dfrac{5i + 3i^2}{2i^2}$

$\dfrac{-3 + 5i}{-2}$

$\dfrac{3}{2} - \dfrac{5}{2}i$

d. $\dfrac{4 - 3i}{2 + 5i} \cdot \dfrac{2 - 5i}{2 - 5i}$

$\dfrac{8 - 20i - 6i + 15i^2}{4 + 25}$

$-\dfrac{7}{29} - \dfrac{26}{29}i$

EXAMPLE 9

Solve $2x^2 + 4x + 20 = 0$ over the complex number system. Write your answers in standard complex form.

Solution

Using the quadratic formula, we get $x = -1 \pm 3i$.

Chapter 8 Review Exercises

1. The profit at Big Jim's Mart the first year it was open can be modeled by

$$P(m) = 10.3 + 2\sqrt{m}$$

where $P(m)$ represents the profit in thousands of dollars for Big Jim's Mart during the mth month of the year.

  a. Find the profit for Big Jim's Mart in the third month of the year.

  b. Find the profit for Big Jim's Mart during August. [8.1]

1. The weight of an alpaca, a grazing animal that is found mostly in the Andes, can be modeled by

$$W(a) = 22.37\sqrt[5]{a^3}$$

where $W(a)$ represents the body weight in kilograms of an alpaca that is a years old.

Source: Model derived from data given in the Alpaca Registry Journal, *Volume II, Summer–Fall 1997.*

© Neil Sutherland/Alamy

  a. Find the body weight of a four-year-old alpaca. [8.1]

  b. Estimate the age of an alpaca that weighs 60 kilograms. [8.4]

2. Give the domain and range of the following functions.

  a. $f(x) = -5\sqrt{x}$

  b. $h(x) = 5\sqrt{x}$

  c. $k(a) = 3\sqrt[4]{a}$

  d. $M(p) = -2.3\sqrt[6]{p}$

  e. $g(x) = 3.4\sqrt[5]{x}$

  f. $f(x) = -2\sqrt[3]{x}$

  g.

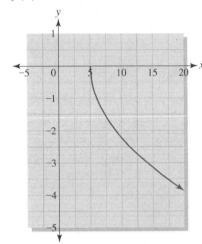

  [8.1]

3. Simplify the following radical expressions.

  a. $\sqrt{144x^2y^4}$ **b.** $\sqrt{100a^3b}$

  c. $\sqrt{180m^4n^5}$ **d.** $\sqrt[3]{64x^3y^5}$

  e. $\sqrt[5]{32a^5b^{10}c^{30}}$ **f.** $\sqrt[4]{25m^3n^{11}}$

  g. $\sqrt[3]{-8x^3y^7}$ **h.** $\sqrt[5]{-40a^7b^3c^5}$ [8.2]

4. Perform the indicated operation and simplify.

  a. $5\sqrt{3x} + 2\sqrt{3x}$

  b. $3x\sqrt{5y} - 6\sqrt{5x^2y}$

  c. $\sqrt[3]{27a^3b^2} - 5a\sqrt[3]{b^2}$

  d. $\sqrt[5]{-64a^7b^3c^{10}} + 4ac\sqrt[5]{2a^2b^3c^5}$
       $+\ 19ac^2\sqrt[5]{2a^2b^3}$

  e. $5x\sqrt{3xy^3} + 7\sqrt{27x^3y^3} + 4\sqrt{3y}$ [82]

5. Multiply the following radical expressions and simplify the result.

  a. $\sqrt{15} \cdot \sqrt{20}$

  b. $\sqrt{3xy} \cdot \sqrt{6x}$

c. $4\sqrt{5a^3b} \cdot 3\sqrt{2ab}$

d. $\sqrt[3]{3a^3b^2} \cdot \sqrt[3]{18b^2}$

e. $3x^2y\sqrt[3]{15xy^2z} \cdot 2xy^3\sqrt[3]{18xyz^5}$

f. $(3 + 5\sqrt{7})(2 - 3\sqrt{7})$

g. $(3 + 5\sqrt{6})(3 - 5\sqrt{6})$ [8.3]

6. Simplify the following radical expressions. Rationalize the denominator if necessary.

a. $\sqrt{\dfrac{36}{2}}$

b. $\dfrac{\sqrt{24}}{\sqrt{3}}$

c. $\dfrac{\sqrt{5x}}{\sqrt{3x}}$

d. $\sqrt{\dfrac{7ab}{3ac}}$

e. $\dfrac{5x}{\sqrt[3]{2x^2y}}$

f. $\dfrac{5}{2 + \sqrt{3}}$

g. $\dfrac{5 + 2\sqrt{x}}{3 - 4\sqrt{x}}$

h. $\dfrac{4 - \sqrt{7}}{2 + \sqrt{6}}$ [8.3]

7. Find the period of a pendulum if its length is 2 ft.[8.4]

8. If you want a pendulum to have a period of 2.5 seconds, what must its length be?[8.4]

10. Use the model from Exercise 1 to estimate the month in which the profit for Big Jim's Mart will be $17,000. [8.4]

11. Solve the following radical equations.

a. $\sqrt{5 + x} = 4$

b. $\sqrt{3x - 7} = 2$

c. $-3\sqrt{2x - 7} + 14 = -1$

d. $\sqrt{x - 4} = \sqrt{2x + 8}$

e. $\sqrt{x + 2} = \sqrt{3x - 7}$

f. $\sqrt{x - 5} - \sqrt{x} = -1$

g. $\sqrt{x - 4} = 5 + \sqrt{3x}$

h. $\sqrt{x + 5} + \sqrt{3x + 4} = 13$

i. $\sqrt[3]{2x} = 3$

j. $\sqrt[4]{5x + 2} = 2$ [8.4]

12. Simplify the given expressions and write your answer in terms of i.

a. $\sqrt{-25}$

b. $\sqrt{-32}$

c. $\sqrt{-4} + \sqrt{-25}$

d. $(2 + \sqrt{-5})(3 - \sqrt{-10})$ [8.5]

13. Perform the indicated operation and write your answer in the standard form for a complex number.

a. $(2 + 3i) + (5 + 7i)$

b. $(3.5 + 1.2i) + (2.4 - 3.6i)$

c. $(4 + 15i) - (3 + 11i)$

d. $(4.5 + 2.9i) - (1.6 - 4.2i)$

e. $(3 + 4i)(2 - 7i)$

f. $(6 + 5i)(2 + 7i)$

g. $(2 - 3i)(2 + 3i)$

h. $(2.3 + 4.1i)(3.7 - 9.2i)$

i. $\dfrac{12 + 7i}{3}$

j. $\dfrac{12 + 9i}{3}$

k. $\dfrac{4 + 7i}{5i}$

l. $\dfrac{7}{2 + 3i}$

m. $\dfrac{5 + 7i}{3 - 4i}$

n. $\dfrac{2 - 9i}{4 + 3i}$

o. $\dfrac{2.5 + 6.4i}{3.3 + 8.2i}$

p. $\dfrac{1.5 + 7.25i}{3.25 - 4.5i}$ [8.5]

14. Solve each equation and write any complex solutions in standard form.

a. $x^2 + 25 = 0$

b. $x^2 - 4 = -20$

c. $2(x - 7)^2 + 11 = 5$

d. $3x^2 + 12x + 15 = 2$

e. $-x^2 - 4x + 6 = 15$ [8.5]

Chapter 8 Test

1. The speed of a sound wave in air can be modeled by

$$v(T) = 20.1\sqrt{273 + T}$$

where $v(T)$ represents the speed of sound in air in meters per second when the temperature is T degrees Celsius. Find the speed of a sound wave at room temperature (20 degrees Celsius).

2. A daredevil wants to free-fall from a plane for 1.5 minutes before he needs to pull his ripcord and release his parachute at 1500 feet. How high should the plane be to allow for this long of a free-fall? (See the formula in exercise 2 of section 8.1.)

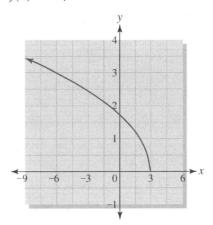

© ImageState/Alamy

3. Give the domain and range of the following functions.

a. $f(x) = -4.5\sqrt{x}$

b. $f(x) = 4\sqrt[3]{x}$

c.

4. Simplify the following radical expressions. Rationalize the denominator if necessary.

a. $\dfrac{\sqrt{6}}{\sqrt{5x}}$

b. $\sqrt{\dfrac{8b}{10a}}$

c. $\dfrac{5m}{\sqrt[3]{3mn^2}}$

d. $\dfrac{3 - \sqrt{2}}{2 + \sqrt{5}}$

5. Simplify the following radical expressions.

a. $\sqrt{36xy^2}$

b. $7\sqrt{120a^2b^3}$

c. $\sqrt[5]{-64m^3n^7p^{10}}$

6. Solve each equation and write any complex solutions in standard form.

a. $-4(x + 5)^2 + 7 = 9$

b. $2.3x^2 + 4.6x + 9 = 5$

7. If you want a pendulum to have a period of 3 seconds, what must its length be?

8. Perform the indicated operation and simplify.

a. $5n\sqrt{6m} - 2\sqrt{24mn^2}$

b. $2a\sqrt{7ab^5} + 7b\sqrt{28a^3b^3} + 4b\sqrt{7a}$

9. Solve the following radical equations.

a. $2\sqrt{5x + 4} - 11 = -3$

b. $\sqrt{x - 13} - \sqrt{x} = -1$

c. $\sqrt{x - 7} + \sqrt{4x - 11} = 13$

d. $\sqrt[3]{x + 5} = 4$

10. Multiply the following radical expressions and simplify the result.

a. $\sqrt{3ab} \cdot 2\sqrt{15ac}$

b. $\sqrt[3]{4a^4b} \cdot \sqrt[3]{18b^2}$

c. $2xy\sqrt[3]{18xz^2} \cdot 5xz^2\sqrt[3]{6x^7yz^4}$

d. $(5 + 2\sqrt{3})(2 - 4\sqrt{3})$

11. Perform the indicated operation and write your answer in the standard form for a complex number.

   a. $(2.7 + 3.4i) + (1.4 - 4.8i)$

   b. $(5 + 11i) - (4 - 7i)$

   c. $(7 + 2i)(3 - 5i)$

   d. $(1.5 - 4.5i)(2.25 - 6.5i)$

   e. $\dfrac{8}{4 + 5i}$

   f. $\dfrac{3 + 2i}{6 - 7i}$

12. In helium the speed of sound can be modeled by

$$v(T) = 58.8\sqrt{273 + T}$$

where $v(T)$ represents the speed of sound in helium in meters per second when the temperature is T degrees Celsius.

   a. Find the speed of a sound wave at room temperature (20 degrees Celsius).

   b. How does the speed of sound in air compare to the speed of sound in helium?

Chapter 8 Projects

Written Project

One or more people

HOW TIGHT CAN THAT TURN BE?

In this project you will investigate the relationship between the speed of a Segway Human Transporter and the radius of a curve on which it is traveling. Although a Segway can "turn on a dime" when stopped, if it is traveling at any speed, it will need a certain turning radius to turn safely. According to the manufacturer, a Segway travels at top speeds ranging from 6 to 12.5 miles per hour depending on the settings you choose. Use this information and the formula

$$v = \sqrt{4.6r}$$

where v is the velocity in miles per hour and r is the minimum turning radius in feet, to answer the following questions.

Write up

a. What is the maximum speed of a Segway on a turn with radius of 25 ft?

b. What turning radius must be available to turn at the maximum speed of 12.5 miles per hour?

c. What turning radius must be available to turn at 6 miles per hour?

d. The turning radius of a bicycle is similar to that of a Segway, but a bicycle can travel at a much higher speed. Find the turning radius of a bicycle traveling at 20 miles per hour.

e. The Tour de France is a 23-day bicycle race that takes place in and around France each year during the month of July. In 2004 the race was about 2100 miles long, and Lance Armstrong finished the 23-day race in about 83 hours and 36 minutes. Use this information to determine the average speed of the race and what turning radius would be needed to turn at that average speed.

f. The next-to-last stage of the Tour de France in 2004 was approximately 34.4 miles long, and Armstrong finished that stage in just under 1 hour and 7 minutes. Find Armstrong's average speed and what turning radius would be needed for him to make a turn at that speed. Do you think he ever made a turn at this speed? Why or why not?

g. If this formula is used to represent the velocity and turning radius of a Segway, what are a reasonable domain and range for your model?

h. If this formula is used to represent the velocity and turning radius of a bicycle, what are a reasonable domain and range for your model?

i. If you gave different domains and ranges in parts g and h, explain why.

Research Project

One or more people

What you will need:

- Find information about the orbital radius and period of different planets and moons.

WHAT IS THAT PLANET'S PERIOD?

In this project you will explore the time it takes different planets to go around the sun. The earth has a period of 1 year, or 365.25 days, and has an orbital radius of about 150 million kilometers. In 1610 the German mathematician and astronomer Johannes Kepler formulated what became known as Kepler's Third Law:

$$T^2 = R^3$$

where T is the period of a planet with an average orbital radius of R. We will use an earth year, 365.25 days, to be a standard unit of time and millions of kilometers for the units for radius. Making these the standard units will give us a variation constant that will work for all planets that orbit our sun. Solving for T, we get

$$T = 0.000544 \sqrt{R^3}$$

Use this formula to answer the following questions.

Write up

a. Mercury has an orbital radius of about 58 million kilometers. Find its period.

b. Mars has an orbital radius of about 228 million kilometers. Find its period.

c. Find the average orbital radius of the other planets. Give the source of your information.

d. Use the values that you found for the other planets to calculate their periods.

e. If you wanted to have a satellite orbit the sun every 1/2 year, how far from the sun must it orbit?

   Use the basic formula $T = k\sqrt{R^3}$ to answer the following questions.

f. Saturn has over 31 known moons. One of Saturn's moons, Hyperion, has an orbital radius of 1.48 million kilometers and a period of about 21.3 days. Use this information to find a model for the orbits of Saturn's moons.

g. Another of Saturn's moons, Titan, has an orbital radius of about 1.22 million kilometers. Find the period of Titan's orbit.

h. Epimetheus orbits Saturn in only 0.694 day. Find its orbital radius.

i. Find the orbital radius of two more of Saturn's moons and calculate their periods.

Research Project

One or more people

What you will need:

- Find a real-world situation that can be modeled with a radical function.
- You might want to use the Internet or library.

EXPLORE YOUR OWN RADICAL FUNCTION

In this project you are given the task of finding and exploring a real-world situation that you can model using a radical function. You may use the problems in this chapter to get ideas of things to investigate, but your application should not be discussed in this textbook. Some items that you might wish to investigate would be formulas from physics and other sciences. Remember that any kind of root can be used, not just square roots.

Write up

a. Describe the real-world situation that you found and where you found it. Cite any sources you used.

b. Either run your own experiment or find data that can be used to verify the relationship that you are investigating.

c. Create a scatterplot of the data on the calculator or computer and print it out or draw it neatly by hand on graph paper.

d. Find a model to fit the data.

e. What is a reasonable domain for your model?

f. Use your model to estimate an output value of your model for an input value that you did not collect in your original data.

g. Use your model to estimate the input value for which your model will give you a specific output value that you did not collect in your original data.